Ergebnisse der Mathematik und ihrer Grenzgebiete

Band 57

Erik M. Alfsen

Compact Convex Sets and Boundary Integrals

With 3 Figures

Springer-Verlag Berlin Heidelberg New York 1971

Erik M. Alfsen

Associate Professor of Mathematics
Department of Mathematics, University of Oslo
Blindern, Oslo 3, Norway

AMS Subject Classifications (1970)

Primary 46-02
Secondary 22D25, 26A51, 31C99, 46A05, 46B99, 46E15, 46G99, 46J10, 46L05, 47D20, 52A05, 54C35, 54C40, 54C45, 62B15

ISBN-13:978-3-642-65011-6 e-ISBN-13:978-3-642-65009-3
DOI: 10.1007/978-3-642-65009-3

 Library of Congress Catalog Card Number 72-136352.
Softcover reprint of the hardcover 1st edition 1971

Preface

The importance of convexity arguments in functional analysis has long been realized, but a comprehensive theory of infinite-dimensional convex sets has hardly existed for more than a decade. In fact, the integral representation theorems of Choquet and Bishop - de Leeuw together with the uniqueness theorem of Choquet inaugurated a new epoch in infinite-dimensional convexity. Initially considered curious and technically difficult, these theorems attracted many mathematicians, and the proofs were gradually simplified and fitted into a general theory. The results can no longer be considered very "deep" or difficult, but they certainly remain all the more important. Today Choquet Theory provides a unified approach to integral representations in fields as diverse as potential theory, probability, function algebras, operator theory, group representations and ergodic theory. At the same time the new concepts and results have made it possible, and relevant, to ask new questions within the abstract theory itself. Such questions pertain to the interplay between compact convex sets K and their associated spaces $A(K)$ of continuous affine functions; to the duality between faces of K and appropriate ideals of $A(K)$; to dominated-extension problems for continuous affine functions on faces; and to direct convex sum decomposition into faces, as well as to integral formulas generalizing such decompositions. These problems are of geometric interest in their own right, but they are primarily suggested by applications, in particular to operator theory and function algebras. In this connection it should be noted that there is a rich supply of information available about such convex sets that can arise as state spaces of C^*-algebras; a fact which has been of no minor importance in the approach to the problems mentioned above. In spite of this, the finer theory of compact convex sets still remains very incomplete with ample room for further research which will hopefully increase our insight into the behaviour of infinite-dimensional convex sets, and in turn feed back to C^*-theory and other fields of application.

The purpose of the present book is twofold; first, to give a general and up to date introduction to Choquet Theory which can be used as a text for graduate students as well as a reference book for mathematicians applying the theory in other fields; and, second, to lead up to current research with regard to the finer theory of compact convex sets.

It has been assumed only that the reader has some basic knowledge of functional analysis and integration theory, as examplified by the various Hahn-Banach extension and separation theorems, the Open Mapping theorem, the Krein-Milman theorem and Milman's (converse) theorem, together with the Riesz representation theorem.

The book has two chapters. Chapter I treats the representation of points by boundary measures, while the question of uniqueness is postponed to Chapter II. The latter is more algebraic and goes into the structure of Choquet simplexes and general compact convex sets.

In order to keep the volume within reasonable bounds, it has been necessary to omit a number of interesting topics which are not directly related to the main scope of the book. Such topics are, e.g., affine selections, intersection properties of cells, symmetric convex sets and classification of Banach spaces, tensor products of compact convex sets, projective limits of compact convex sets and injective limits of $A(K)$-spaces, as well as infinite dimensional polytopes. Another sample consists of weakly complete cones, spaces with complex scalars, Bishop's peak point theorem for function algebras, as well as results based on several complex variables such as Rossi's local peak point theorem. Also we have refrained from generality in certain cases where it would lead to technical complications without invoking any essentially new ideas. For example, we have not studied convex sets with a distinguished extreme point and the associated $A_0(K)$-spaces (without unit), nor have we studied well-capped cones.

I am greatly indebted to Christian Skau who helped to improve the manuscript and also to Tage Bai Andersen for many stimulating discussions on the subject of the present book.

Erik M. Alfsen

Oslo, January 1971

Contents

Chapter I

Representations of Points by Boundary Measures

Chapter II

Structure of Compact Convex Sets

Short List of Symbols and Notations

$f<g$	means $f(x)<g(x)$ for *all* x in the common domain of f and g. $\\|f\\|=\\|f\\|_X=\sup_{x\in X}\|f(x)\|$ (uniform norm).
1_X	denotes the (constant) function which takes the value 1 everywhere on X.
$\operatorname{lin} B$	denotes the *linear span* of a set B in a linear space.
$\operatorname{conv} B$	denotes the *convex hull* of a set B in a linear space.
cl. conv. $B=\overline{\operatorname{conv} B}$	denotes the *closed convex hull* of a set B in a topological linear space.
face B = face B	denotes the *face* generated by a subset B of a convex set K in a linear space.
cl. face B	denotes the *closed face* generated by a subset B of a convex set in a topological linear space. (Note that cl. face $B\neq\overline{\text{face } B}$ in general).
$\partial_e K$	denotes the *extreme boundary*, i.e. the set of extreme points, of a convex set K in a linear space.
The vague topology (of measures)	is the w^*-topology of $M_{\mathbb{R}}(X)$ in the duality of $M_{\mathbb{R}}(X)$ and $C_{\mathbb{R}}(X)$.

Chapter I

Representation of Points by Boundary Measures

§ 1. Distinguished Classes of Functions on a Compact Convex Set

Throughout this chapter we shall consider an arbitrary, but fixed, locally convex Hausdorff space E over $\mathbb{R}$. If K and K' are convex subsets of E and $K \subset K'$, then $A(K;K')$ shall denote the vector space of all restrictions to K of continuous affine real-valued functions on K'. For simplicity we write $A(K)$ in the place of $A(K;K)$, and we note that generally

$$A(K;E) \subset A(K;K') \subset A(K). \tag{1.1}$$

Note also that $A(E) = \mathbb{R} \oplus E^*$, since every continuous affine function a on E can be decomposed as $a = a(0) + a'$ where a' is a continuous linear function on E.

In the sequel we shall restrict ourselves to the case where K is compact, and we note that $A(K)$ is a uniformly closed subspace of $C_{\mathbb{R}}(K)$.

The convex cone of all continuous and convex real-valued functions on K will be denoted by $P(K)$, and the convex cone of all lower semicontinuous convex functions on K with values in $]-\infty, \infty]$ will be denoted by $Q(K)$.

Proposition I.1.1. *If K is a compact convex set, then $P(K) - P(K)$ is a uniformly dense vector subspace of $C_{\mathbb{R}}(K)$.*

Proof. Clearly $A(K;E) \subset P(K) - P(K)$, and $A(K;E)$ separates the points of K (by the Hahn-Banach Theorem). Now the density follows by Stone's Theorem, since $P(K)$ is a sup-closed cone, and consequently $P(K) - P(K)$ is a lattice subspace of $C_{\mathbb{R}}(K)$.

In fact if $f_i = h_i - k_i$ with $h_i, k_i \in P(K)$ for $i = 1, 2$, then the functions $f_i' = f_i + (k_1 + k_2)$ belong to $P(K)$ for $i = 1, 2$, and so

$$\begin{aligned} f_1 \vee f_2 &= [f_1' - (k_1 + k_2)] \vee [f_2' - (k_1 + k_2)] \\ &= f_1' \vee f_2' - (k_1 + k_2) \in P(K) - P(K). \end{aligned}$$

This completes the proof. $\square$

Clearly $P(K)$ and $Q(K)$ are both closed under *finite suprema;* $P(K)$ is closed under *uniform limits* of increasing nets, and $Q(K)$ is closed under *pointwise limits* of increasing nets. We shall see that $P(K)$ and $Q(K)$ are the smallest extensions of $A(K)$, with these properties. (Corollary I.1.3 below.)

Proposition I.1.2. *If K is a compact convex set and $f \in Q(K)$, then for every $x \in K$:*

$$f(x) = \sup\{a(x) \mid f > a \in A(K; E)\}. \tag{1.2}$$

Proof. Let $M \subset E \times \mathbb{R}$ be the *supergraph* of f, i.e. $(y, \alpha) \in M$ iff $y \in K$ and $\alpha \geqq f(y)$. We observe that lower semi-continuity of f is equivalent to closedness of M, and that convexity of f is equivalent to convexity of M.

Let x be some fixed element of K, and let $\beta < f(x)$ be arbitrary. We shall prove that there exists an $a \in A(K; E)$ such that $a < f$ and $\beta < a(x)$.

Case 1: $f(x) < \infty$. By Hahn-Banach separation there exists a closed hyperplane H in $E \times \mathbb{R}$ which separates M strictly from the point (x, β).

We know that a hyperplane in $E \times \mathbb{R}$ is the graph of an affine function unless it is of the form $H_1 \times \mathbb{R}$ where H_1 is a hyperplane in E, which is impossible in the present situation with H separating (x, β) strictly from $(x, f(x))$. Hence H is the graph of an affine function $a: E \to \mathbb{R}$, and the open half-spaces associated with H are $\{(y, \alpha) \mid \alpha < a(y)\}$ and $\{(y, \alpha) \mid \alpha > a(y)\}$. By assumption one of these contains (x, β), the other all of M. Hence we get:

$$\beta < a(x), \qquad a < f.$$

Finally $a^{-1}(0)$ is closed since $a^{-1}(0) \times \{0\} = H \cap (E \times \{0\})$. Hence a is continuous.

Case 2: $f(x) = \infty$. Let β' be an arbitrary number such that $\beta < \beta' < \infty$, and define

$$M' = \mathrm{cl.conv}\,((x, \beta') \cup M).$$

We claim that $(x, \beta) \notin M'$.
To prove this claim, we define two auxiliary sets:

$$A = \{(y, \alpha) \mid y \in K, \alpha \geqq \beta'\},$$
$$B = \{(y, \alpha) \mid y \in K, f(y) \leqq \alpha \leqq \beta'\}.$$

Every point of $\mathrm{conv}\,((x, \beta') \cup M)$ must belong to A or to $\mathrm{conv}\,((x, \beta') \cup B)$. Hence

$$M' \subset A \cup \mathrm{cl.conv}\,((x, \beta') \cup B).$$

By an elementary theorem on the convex hull of a finite union of compact sets, one may omit the prefix "cl" in this formula. Clearly

$$(x,\beta)\notin A\cup \operatorname{conv}((x,\beta')\cup B),$$

and so we have verified that $(x,\beta)\notin M'$.

From now on one may argue as in the first case with M' in place of M. □

Corollary I.1.3. *If K is a compact convex set, then $Q(K)$ consists of all pointwise limits of increasing nets of functions of the form $a_1\vee\cdots\vee a_n$ where $a_i\in A(K;E)$ for $i=1,\dots,n$. Similarly $P(K)$ consists of all uniform limits of increasing nets of such functions.*

Proof. Application of Proposition I.1.2 and of Dini's Lemma. □

Corollary I.1.4. *If K is a compact convex set and $a: K\to]-\infty,\infty]$ is a l. s. c. affine function, then there is an increasing net from $A(K;E)$ converging pointwise to a.*

Proof. By Proposition I.1.2 it is sufficient to prove that the set

$$\{a' \mid a>a'\in A(K;E)\}$$

is directed upward ("filtering to the right"). To this end we assume that $a_1,a_2\in A(K;E)$ and $a_1,a_2<a$. Without lack of generality we may also assume that all functions a, a_1, a_2 are positive.

Again we consider the *supergraph* M of a, and we also consider the *positive subgraphs*.

$$M_i=\{(y,\alpha)\mid y\in K, 0\leqq\alpha\leqq a_i(y)\},\qquad i=1,2.$$

The assumption $a_1,a_2<a$ entails

$$M\cap(M_1\cup M_2)=\emptyset,$$

and by the affinity of a

$$M\cap\operatorname{conv}(M_1\cup M_2)=\emptyset.$$

The set $\operatorname{conv}(M_1\cup M_2)$ is compact by the compactness of M_1 and M_2. Hence we may proceed as in the proof of Proposition I.1.2 to obtain a closed hyperplane H in $E\pm\mathbb{R}$ which separates M from $\operatorname{conv}(M_1\cup M_2)$, and which is the graph of a continuous affine function $a': E\to\mathbb{R}$ such that

$$a_1,a_2<a'<a.$$

This completes the proof. □

Corollary I.1.5. *If K is a compact convex set and $a \in A(K)$, then there is an increasing sequence from $A(K;E)$ converging uniformly to a. In particular $A(K;E)$ is dense in $A(K)$ with respect to the uniform norm.*

Proof. Application of Corollary I.1.4 and of Dini's Lemma. □

Remark: If one is interested only in the density statement of the above Corollary, then one may consider an arbitrary $a \in A(K)$ and separate graph(a) from graph($a-\varepsilon$) by the graph of some continuous affine function $a' : E \to \mathbb{R}$.

We shall denote the *extreme boundary* (i.e. the set of extreme points) of a convex set K by the symbol $\partial_e K$. Let K be convex compact, and consider a function $f: X \to [\alpha, \infty]$ where $\partial_e K \subset X \subset K$ and $\alpha \in \mathbb{R}$. The *lower envelope* of f is the function

$$\check{f} = \sup\{a \mid a \in A(K;E), a|X \leqq f\}. \tag{1.3}$$

Similarly if $f: X \to [-\infty, \alpha]$ where $\partial_e K \subset X \subset K$ and $\alpha \in \mathbb{R}$, then the *upper envelope* of f is defined to be the function

$$\hat{f} = \inf\{a \mid a \in A(K;E), a|X \geqq f\}. \tag{1.4}$$

Note that $\check{f}$ and $\hat{f}$ are defined on the whole set K and not only on X.

Clearly $\check{f} \in Q(K)$. In fact it follows from Proposition I.1.2 that $\check{f}$ is the greatest l.s.c. convex function on K which minorizes f on its domain X. Also is follows that the class $A(K;E)$ occuring in (1.3) may be replaced by any of the larger classes $A(K)$, $P(K)$ or $Q(K)$, and that the sign $\leqq$ may be replaced by $<$.

Dually $\hat{f} \in -Q(K)$, and $\hat{f}$ is the smallest u.s.c. concave function on K which majorizes f on X. Also the class $A(K;E)$ of (1.4) may be replaced by $A(K)$, $-P(K)$, or $-Q(K)$, and the sign $\geqq$ may be replaced by $>$.

Proposition I.1.6. *Let K be a compact convex set, let $\partial_e K \subset X \subset K$ and consider functions $f, g: X \to [-\infty, \alpha]$ where $\alpha \in \mathbb{R}$. Then we shall have*

$$\hat{f}(x) \leqq \alpha, \quad \text{all } x \in K, \tag{1.5}$$

$$f \leqq g \quad \text{on } X \Rightarrow \hat{f} \leqq \hat{g}, \tag{1.6}$$

$$\widehat{f+g} \leqq \hat{f} + \hat{g}, \tag{1.7}$$

$$\widehat{\alpha f} = \alpha \hat{f}, \quad \text{all } \alpha \geqq 0, \tag{1.8}$$

$$\widecheck{(-f)} = -\hat{f}. \tag{1.9}$$

If $f,g: X\to[\alpha,\infty]$, then the lower envelopes will enjoy the corresponding dual properties. Finally if f is a bounded real valued function on X, then $\hat{f}$ and $\check{f}$ are both defined, and

$$\check{f}(x)\leqq\hat{f}(x),\quad \text{all } x\in K. \tag{1.10}$$

The *proofs* are all straightforward, except perhaps for the verification of (1.10) which is based on the Krein-Milman Theorem. □

* We shall apply Corollary I.1.5 to obtain Grothendieck's Completeness Theorem, and then pass to the proof of the Krein-Šmullyan Theorem which will be needed in Ch. II, § 2.

Let $\mathscr{A}$ denote the collection of all convex, balanced, w^*-closed, and equicontinuous subsets of E^*. Thus $\mathscr{A}$ consists of all polars of neighbourhoods of the origin in E. Also let $\bar{E}$ denote the vector space of all linear functionals on E^* with w^*-continuous restrictions to the sets in $\mathscr{A}$, provided with the (locally convex, Hausdorff) topology of uniform convergence on the sets in $\mathscr{A}$.

Lemma I.1.7. *Every $\varphi\in\bar{E}$ can be approximated in the topology of $\bar{E}$ by evaluations $\tilde{x}: f\rightsquigarrow f(x)$, where $x\in E$ and $f\in E^*$.*

Proof. Consider a neighbourhood of the origin in $\bar{E}$, say

$$V(A;\varepsilon)=\{\psi\in\bar{E}\,|\,|\psi(f)|<\varepsilon \text{ all } f\in A\}, \tag{1.11}$$

where $A\in\mathscr{A}$, $\varepsilon>0$.

The set A is w^*-compact (cf. e.g. [86], [240]), and by Corollary I.1.5 there is a w^*-continuous *affine* function a on E^* such that

$$|\varphi(f)-a(f)|<\frac{\varepsilon}{2},\quad \text{all } f\in A. \tag{1.12}$$

The function $f\rightsquigarrow a(f)-a(0)$ is a w^*-continuous *linear* function on E^*, and by an elementary theorem it is the evalution at some point $x\in E$ (cf. e. g. [86], [240]). By virtue of (1.12) and since $0\in A$, we obtain

$$|\varphi(f)-f(x)|\leqq|\varphi(f)-a(f)|+|a(0)|<\varepsilon,$$

for every $f\in A$. Hence $\varphi-\tilde{x}\in V(A;\varepsilon)$, and the proof is complete. □

Theorem I.1.8. *(Grothendieck) The space $\bar{E}$ is (canonically isomorphic and homeomorphic to) the completion of E. In particular, E is complete iff every member of $\bar{E}$ is the evaluation at some point of E.*

Proof. 1. Let $\{\varphi_\alpha\}$ be a Cauchy-net from $\bar{E}$ with pointwise limit φ. By the definition of the topology of $\bar{E}$, the convergence is uniform on every $A\in\mathscr{A}$. Hence $\varphi\in\bar{E}$, and so $\bar{E}$ is *complete.*

By Lemma I.1.7, the canonical mapping $x\rightsquigarrow\tilde{x}$ is a linear isomorphism of E onto a *dense subset* of $\bar{E}$.

A fundamental system of neighbourhoods of the origin in $\bar{E}$ consists of the polars of the sets $A\in\mathscr{A}$. Thus it follows by the Bi-Polar Theorem that $x\rightsquigarrow\tilde{x}$ is a homeomorphic embedding of E into $\bar{E}$. This completes the proof. □

Corollary I.1.9. *If E is a Banach space and φ is a linear functional on E^* which has a w^*-continuous restriction to the unit ball $E_1^*=\{f\in E^*\mid \|f\|\leqq 1\}$, then φ is the evalutation at some point of E.*

The *proof* is obvious since a subset of E^* is equicontinuous iff it is contained in some scalar multiple of E_1^*. □

Lemma I.1.10. *Let W be a subset of E^* such that $A\cap W$ is w^*-open in A for every $A\in\mathscr{A}$. Assume that U, V are neighbourhoods of 0 in E and that I is some subset of E such that*

$$I^0\cap U^0\subset W. \tag{1.13}$$

Then there exists a finite subset J of U such that

$$(I\cup J)^0\cap V^0\subset W. \tag{1.14}$$

Proof. Define for every $x\in E$:

$$B_x=V^0\cap I^0\cap\{x\}^0.$$

Clearly B_x is a closed subset of the w^*-compact set V^0, and

$$\bigcap_{x\in U} B_x=V^0\cap I^0\cap U^0\subset V^0\cap W.$$

Now $V^0\in\mathscr{A}$, and by assumption $V^0\cap W$ is w^*-open in V^0. By a well known compactness argument there is a *finite* subset J of U such that

$$\bigcap_{x\in J} B_x\subset V^0\cap W.$$

By the definition of B_x we obtain (1.14). □

Proposition I.1.11. *If E is metrizable, then there is a locally convex (Hausdorff) topology $\mathscr{T}'$ on E^* such that a subset F of E^* is $\mathscr{T}'$-closed iff $F\cap A$ is w^*-closed for every $A\in\mathscr{A}$. Specifically, $\mathscr{T}'$ is the topology of uniform convergence on sequences converging to 0.*

Proof. 1. Let $F \subset E^*$ be closed in the topology $\mathscr{T}'$ of uniform convergence on sequences converging to 0. We consider an arbitrary $A \in \mathscr{A}$, and we shall prove that $A \cap F$ is w^*-closed; or what is equivalent, that $A \backslash F$ is relatively w^*-open in A.

For any $f \in A \backslash F$ there exists a sequence $\{x_n\}$ converging to 0 in E, such that the corresponding $\mathscr{T}'$-neighbourhood

$$W = \{g \in E^* \,|\, |g(x_n)| \leqq 1, \; n = 1, 2, \ldots\}.$$

satisfies

$$(f + W) \cap F = \emptyset. \tag{1.15}$$

By definition $A = U^0$ for some neighbourhood U of 0 in E, and by the continuity of f there is a neighbourhood V of 0 in E such that $|f(x)| < \frac{1}{2}$ for $x \in V$. Now we choose n_0 such that $x_n \in \frac{1}{2} U \cap V$ for $n > n_0$, and consider the w^*-neighbourhood

$$W' = \{g \in E^* \,|\, |g(x_n)| \leqq 1, \; n = 1, \ldots, n_0\}.$$

We observe that

$$(f + W') \cap A \subset f + W,$$

and it follows by (1.15) that f is an interior point of $A \backslash F$ in the relative w^*-topology of A.

2. We assume that $A \cap F$ is w^*-closed for every $A \in \mathscr{A}$, and we shall prove that F is $\mathscr{T}'$-closed. We consider an arbitrary $f \in \complement F$, and we shall prove that the set $W = \complement F - f$ contains a typical $\mathscr{T}'$-neighbourhood of 0

$$\{g \in E^* \,|\, |g(x_n)| \leqq 1, \; n = 1, 2, \ldots\}, \tag{1.16}$$

where $\{x_n\}$ converges to 0 in E.

We observe that $(F - f) \cap A$ is w^*-closed for every $A \in \mathscr{A}$, and so $W \cap A$ is w^*-open in A for every $A \in \mathscr{A}$. By metrizability there is a decreasing sequence $\{U_n\}_{n=1}$ of neighbourhoods of 0 in E which form a fundamental system ("a local base"). Since $W \cap U_1^0$ is w^*-open in U_1^0 and $0 \in W \cap U_1^0$, we may apply the definition of the w^*-topology to yield a finite subset I_1 of E such that

$$I_1^0 \cap U_1^0 \subset W.$$

By Lemma I.1.10 there is a finite subset I_2 of U_1 such that

$$(I_1 \cup I_2)^0 \cap U_2^0 \subset W.$$

Repeating this process inductively, one can choose a sequence $\{I_n\}$ of finite sets such that $I_{n+1} \subset U_n$ and

$$(I_1 \cup \cdots \cup I_n)^0 \cap U_n^0 \subset W. \tag{1.17}$$

Clearly the elements of the set $S=\bigcup_{n=1}^{\infty} I_n$ can be arranged to a sequence converging to 0, and the corresponding $\mathscr{T}'$-neighbourhood (1.16) is equal to S^0. By virtue of (1.17) and the fact that $\bigcup_{n=1}^{\infty} U_n^0=E^*$

$$S^0=\bigcup_{n=1}^{\infty}(S^0\cap U_n^0)\subset W,$$

and the proof is complete. □

Remark. Clearly the topology $\mathscr{T}'$ of Proposition I.1.11 is the finest (strongest) topology on E^* which agrees with the w^*-topology on equicontinuous sets. It is easily verified that it can be described also as the topology of uniform convergence on *compact* sets, and on *totally bounded* (precompact) sets (cf. e. g. [240]).

Theorem I.1.12. *(Krein-Šmullyan) If E is a Frechet space and F is a convex subset of E^* such that $F\cap A$ is w^*-closed for every $A\in\mathscr{A}$, then F is w^*-closed.*

Proof. We consider an element f in $\complement F$. By Lemma I.1.11, F is $\mathscr{T}'$-closed, and so there exists a $\mathscr{T}'$-continuous linear functional φ on E^* which separates f strictly from F.

Now consider a set A in $\mathscr{A}$ and let $g\in A$ and $\varepsilon>0$. The set

$$W=\{h\in E^*|\ |\varphi(h)-\varphi(g)|<\varepsilon\},$$

is $\mathscr{T}'$-open, and by Lemma I.1.11, the set $A\cap W$ is a w^*-neighbourhood of g in A. Hence $\varphi|A$ is w^*-continuous, and it follows from Theorem I.1.8 that φ is w^*-continuous. Since φ separates f strictly from F, the set $\complement F$ is a w^*-neighbourhood of f, and the proof is complete. □

Corollary I.1.13. *If E is a Banach space and F is a convex subset of E^* such that $F\cap E_\rho^*$ is w^*-closed for every ball $E_\rho^*=\{f\in E^*|\ \|f\|\leqq\rho\}$ with $\rho>0$, then F is w^*-closed.*

Notes. The first, and essential, step towards the results of the preceding "starred" section was made by S. Banach in his famous book of 1932 [34]. He proved Corollary I.1.13 for the case of a vector subspace F of a Banach dual space E^*. In his terminology it reads: "Every transfinitely closed vector subspace F of E^* is regularly closed" ("regularly closed" ≡ w^*-closed, and "transfinitely closed" ≡ w^*-closed intersection with every E_ρ^* [34]).—The formulation given in Corollary I.1.13 for vector subspaces F of E^* first appeared in a note by N. Bourbaki in 1938 [83], and it was pointed out by J. Dieudonné in 1942 how this theorem was a restatement of Banach's theorem [131].—The equivalence of the various types of closure mentioned above, was also made explicit at about the same time by L. Alaoglu [4] and by S. Kakutani [233].—The generalization from vector subspaces to convex sets was made by M. Krein und V. Šmullyan in 1940. They proved that a convex subset of a Banach dual space is w^*-closed iff it is "transfinitely closed" [254].—In 1948 J. Dieudonné introduced the "bounded w^*-topology" on a Banach dual space E^*, i.e. the topology whose closed sets are exactly those with a w^*-closed intersection with every E_ρ^*. He showed that this topology coincides with the topology of uniform convergence on compact subsets of E, and the idea of the proof of Proposition I.1.11 goes back to his work [132].—The final step towards Theorem I.1.12 was made in 1949 by J. Dieudonné and L. Schwartz, extending the previous results from Banach spaces to Frechet spaces, and in fact also to inductive limits of Frechet spaces [134].—Theorem I.1.8 was proved by A. Grothendieck in 1950 [204].—An extensive survey of recent generalizations of the Krein-Šmullyan Theorem and related matters can be found in T. Husain's book [217].

§ 2. Weak Integrals, Moments and Barycenters

In the present chapter we shall use the word *measure* to denote a real *Radon measure* μ on some compact Hausdorff space T (i.e. $\mu \in C_{\mathbb{R}}(T)^*$). We recall that the Radon measures on T correspond biuniquely to the *Baire-measures* on T, i.e. to the (real-valued) countably additive set-functions μ_0 on the σ-field $\mathscr{B}_0$ of Baire subsets of T. ($\mathscr{B}_0$ is generated by the "zero sets" $f^{-1}(0)$ with $f \in C_{\mathbb{R}}(T)$, which are exactly the closed G_δ-subsets of T.) Specifically, the correspondence is given by

$$\mu(f) = \int f \, d\mu_0, \quad \text{all } f \in C_{\mathbb{R}}(T). \tag{2.1}$$

Recall also that every Baire measure μ_0 is *regular* in the sense that for every $E \in \mathscr{B}_0$:

$$\begin{aligned} \mu_0(E) &= \sup\{\mu_0(F) \mid E \supset F, F \text{ is closed}\} \\ &= \inf\{\mu_0(U) \mid E \subset U, U \text{ is open}\}. \end{aligned} \tag{2.2}$$

Moreover, every Baire measure μ_0 can be uniquely extended to a unique *regular Borel measure* $\tilde{\mu}_0$, i.e. to a countably additive set function satisfying the condition (2.2) on the σ-field $\mathscr{B}$ of Borel subsets of T. ($\mathscr{B}$ is generated by all closed subsets of T.)

Note that every closed (open) Baire set is a G_δ-set (an F_σ-set). Hence in the formula (2.2) for the Baire measure μ_0, no restriction is imposed by assuming F to be a closed G_δ-set (or U to be an open F_σ-set). The corresponding statement for $\tilde{\mu}_0$ is incorrect.

In §4 of the present chapter we shall have to study (non-regular) extensions of μ_0 to σ-fields other than $\mathscr{B}$. Elsewhere we shall be concerned with the regular Borel extension only, and we shall find it convenient to write μ in the place of $\tilde{\mu}_0$ when no confusion is likely to arise. Accordingly we shall use the expressions $\mu(f)$ and $\int f\,d\mu$ interchangeably to denote integrals with respect to $\tilde{\mu}_0$, and we shall use the term *μ-integrable* to mean integrable with respect to $\tilde{\mu}_0$.

In particular, we shall often have to integrate a bounded semicountinuous function f; being (Borel-) measurable it has an integral with respect to $\tilde{\mu}_0$, which will be denoted by $\mu(f)$. Also we shall say that a property holds *almost everywhere* μ if the set of points for which it fails, are all contained in some Borel set of $\tilde{\mu}_0$-measure zero.

For later references we note that if $\{f_\alpha\}$ is an increasing net from $C_{\mathbb{R}}(T)$, then $f=\sup_\alpha f_\alpha$ is l.s.c., hence (Borel) measurable, and if $\sup_\alpha \mu(f_\alpha)<\infty$ for some positive measure μ, then

$$\mu(f)=\sup_\alpha \mu(f_\alpha)=\lim_\alpha \mu(f_\alpha). \tag{2.3}$$

The corresponding statement subsists if $\{f_\alpha\}$ is allowed to be a net of l. s. c. functions (cf. e. g. [87]).

We shall use the symbol $M_{\mathbb{R}}(T)$ to denote the vector space of (real) measures on T and the symbol $M_1^+(T)$ to denote the convex subset of positive and normalized measures (probability measures). Recall that $M_{\mathbb{R}}(T)$ is the *dual* Banach space of $C_{\mathbb{R}}(T)$, and that the (dual) norm is given by $\|\mu\|=|\mu|(K)$. Also $M_{\mathbb{R}}(T)$ is an (order) *complete vector-lattice.* (Completeness is trivial, and the lattice-property follows by the Jordan decomposition.) Moreover, $M_1^+(T)$ is a vaguely (or w^*) compact, convex subset of $M_{\mathbb{R}}(T)$.

One can easily extend integration theory to functions f with values in $\mathbb{R}^n$ by independent treatment of the component functions $p^i\circ f$, where $p^i((\xi_1,\dots,\xi_n))=\xi_i$ for $i=1,\dots,n$. The functionals $p^1,\dots,p^n$ form a base of $(\mathbb{R}^n)^*$, and so one may generalize this type of vector integration to infinite dimensional spaces E by independent treatment of the functions $p\circ f$ where $p\in E^*$.

Definition: Let μ be a measure on a compact Hausdorff space T. A function f from T into a locally convex Hausdorff space E is *weakly μ-integrable* with *μ-integral* $y\in E$ iff $p\circ f$ is μ-integrable with

$$p(y)=\int (p\circ f)\,d\mu, \tag{2.4}$$

for all $p\in E^*$.

Clearly, weak integrals are unique whenever they exist, and we shall denote them by the customary integral sign, i. e. we shall write $y=\int f\,d\mu$.

Proposition I.2.1. *If f is a continuous function from a compact Hausdorff space T into a locally convex Hausdorff space E such that*

$f(T)\subset K$ where K is some weakly compact convex subset of E, then f is weakly μ-integrable for every measure μ on T. If $\mu\in M_1^+(T)$, then the weak μ-integral of f is in K.

Proof. 1. We first assume that $\mu\in M_1^+(T)$. Define for $p\in E^*$:

$$A_p=\{y\in K \mid p(y)=\mu(p\circ f)\}.$$

Clearly $\{A_p \mid p\in E^*\}$ is a collection of weakly closed subsets of K. We claim that it has finite intersection property.

Let $p_1,\dots,p_n\in E^*$, and define the continuous linear map $\Phi: E\to R^n$ by

$$\Phi(y)=(p_1(y),\dots,p_n(y)),\qquad y\in E.$$

Observe that $A_{p_1}\cap\cdots\cap A_{p_n}\neq\emptyset$ iff

$$(\mu(p_1\circ f),\dots,\mu(p_n\circ f))\in\Phi(K). \tag{2.5}$$

If (2.5) fails, then there is a linear functional ψ on R^n, say $\psi((\xi_1,\dots,\xi_n)) = \sum_{i=1}^n \alpha_i\xi_i$, such that

$$\sup_{u\in\Phi(K)}\psi(u)<\psi(v),$$

where $v=(\mu(p_1\circ f),\dots,\mu(p_n\circ f))$.

Let $q=\sum_{i=1}^n \alpha_i p_i\in E^*$. Then

$$\begin{aligned}\psi(v)&=\sum_{i=1}^n \alpha_i\mu(p_i\circ f)=\mu(q\circ f)\\ &\leqq \sup_{t\in T}(q\circ f(t))\leqq\sup_{y\in K}q(y)\leqq\sup_{u\in\Phi(K)}\psi(u).\end{aligned}$$

This is a contradiction, and the claim is proved.

By compactness there exists a point $y\in\bigcap\{A_p \mid p\in E^*\}$, and the first part of the proof is complete.

2. If μ is an arbitrary measure on T, then by Jordan decomposition $\mu=\alpha_1\mu_1-\alpha_2\mu_2$, where $\mu_1,\mu_2\in M_1^+(T)$, and $\alpha_1,\alpha_2\geqq 0$. By the first part of the proof, f has weak integrals y_1 and y_2 with respect to μ_1 and μ_2, respectively. Now it is easily verified that $\alpha_1 y_1-\alpha_2 y_2$ is the weak integral of f with respect to μ. This completes the proof. □

If μ is a measure on a compact convex subset K of a locally convex Hausdorff space, then the identity function $x: K\to E$ is weakly μ-integrable, and the weak μ-integral

$$\int x\,d\mu(x) \tag{2.6}$$

is called the *moment* of μ. (By common abuse of language we denote "function" and "function value" by the same symbol when no confusion is likely to arise.)

If μ is a positive non-zero measure on K, then the point

$$x_\mu = \mu(K)^{-1} \int x \, d\mu(x), \tag{2.7}$$

is called the *barycenter* of μ. It is in K since $\mu(K)^{-1}\mu$ is in $M_1^+(K)$.

In the sequel we shall most often work with positive and normalized measures, and for these the concepts of moment and barycenter coalesce.

If μ is a measure on K such that

$$\mu(K) = 0, \qquad \int x \, d\mu(x) = 0, \tag{2.8}$$

then μ is said to be a (generalized) *affine dependence* on K. The affine dependences on K form a subspace of $M_{\mathbb{R}}(K)$, which we shall denote by the symbol $N(K)$.

It follows easily from the definition of moment that if μ is a *simple* measure on K, say $\mu = \sum_{i=1}^{n} \alpha_i \varepsilon_{x_i}$, then the moment of μ is the corresponding linear combination, i. e.

$$\int x \, d\mu(x) = \sum_{i=1}^{n} \alpha_i x_i. \tag{2.9}$$

If in addition $\mu(K) = \sum_{i=1}^{n} \alpha_i = \alpha \neq 0$ and $\alpha_i \geqq 0$, then

$$x_\mu = \sum_{i=1}^{n} \alpha^{-1} \alpha_i x_i,$$

and if also $\mu(K) = \sum_{i=1}^{n} \alpha_i = 1$, then the barycenter of μ is simply the convex combination

$$x_\mu = \sum_{i=1}^{n} \alpha_i x_i.$$

Similarly it can be seen that a simple measure $\mu = \sum_{i=1}^{n} \alpha_i \varepsilon_{x_i}$ on K is an affine dependence iff

$$\sum_{i=1}^{n} \alpha_i = 0, \qquad \sum_{i=1}^{n} \alpha_i x_i = 0.$$

Proposition I.2.2. *Let K be a compact convex subset of a locally convex Hausdorff space E. A non-zero positive measure μ on K satisfies*

$$a(x_\mu) = \mu(K)^{-1} \int a(x)\,d\mu(x), \quad \text{all } a \in A(K), \tag{2.10}$$

and

$$f(x_\mu) \leqq \mu(K)^{-1} \int f(x)\,d\mu(x), \quad \text{all } f \in Q(K). \tag{2.11}$$

A non-zero (signed) measure μ on K is an affine dependence iff:

$$\int a(x)\,d\mu(x) = 0, \quad \text{all } a \in A(K). \tag{2.12}$$

Proof. 1. By the definition of barycenter

$$p(x_\mu) = \mu(K)^{-1} \int p(x)\,d\mu(x), \quad \text{all } p \in E^*.$$

Every $a \in A(K;E)$ is of the form $a = p|K + \alpha$ where $p \in E^*$, $\alpha \in \mathbb{R}$. It follows that

$$a(x_\mu) = \mu(K)^{-1} \int p(x)\,d\mu(x) + \alpha = \mu(K)^{-1} \int a(x)\,d\mu(x).$$

By Corollary I.1.5, $A(K;E)$ is uniformly dense in $A(K)$. Hence (2.10) follows.

2. For any $f \in Q(K)$ and any $\varepsilon > 0$ we apply Proposition I.1.2 to construct an $a \in A(K;E)$ such that

$$a \leqq f, \quad a(x_\mu) > f(x_\mu) - \varepsilon.$$

By the first part of the proof:

$$\begin{aligned} f(x_\mu) < a(x_\mu) + \varepsilon &= \mu(K)^{-1} \int a(x)\,d\mu(x) + \varepsilon \\ &\leqq \mu(K)^{-1} \int f(x)\,d\mu(x) + \varepsilon. \end{aligned}$$

This gives (2.11), since $\varepsilon > 0$ was arbitrary.

3. The verification of (2.12) is an application of the density of $A(K;E)$ in $A(K)$, as in the first part of the proof. □

If $\mu \in M_1^+(K)$, then (2.10) reduces to

$$a(x_\mu) = \int a(x)\,d\mu(x), \tag{2.13}$$

for all $a \in A(K)$. Formula (2.13) is sometimes referred to as the "barycenter formula".

By (2.12), $N(K)$ is the *annihilator* of $A(K)$ in the duality of $C_{\mathbb{R}}(K)$ and $M_{\mathbb{R}}(K)$.

Proposition I.2.3. *Let K be a compact convex set in a locally convex Hausdorff space E. Then every measure $\mu \in M_1^+(K)$ can be vaguely approximated by simple measures in $M_1^+(K)$ with the same barycenter as μ.*

Proof. Consider a vague neighbourhood of μ consisting of all $\nu\in M_{\mathbb{R}}(K)$ such that

$$|\mu(f_i)-\nu(f_i)|\leqq\varepsilon, \qquad i=1,\dots,n. \tag{2.14}$$

By uniform continuity there is a closed convex neighbourhood W of 0 in E such that

$$|f_i(x)-f_i(y)|<\varepsilon, \qquad i=1,\dots,n, \tag{2.15}$$

whenever $x,y\in K$ and $x-y\in W$.

Choose a closed convex and balanced neighbourhood W_0 of 0 in E such that $W_0+W_0\subset W$. By compactness there are points $x'_1,\dots,x'_m\in E$ such that $K\subset\bigcup_{j=1}^{m}W_j$ where $W_j=x'_j+W_0$ for $j=1,\dots,m$.

Now define:

$$A_j=K\cap(W_j\setminus(W_1\cup\cdots\cup W_{j-1})), \qquad j=1,\dots,m.$$

Let J be the set of indices for which $\mu(A_j)\neq 0$. Let $\lambda_j=\mu(A_j)$ for $j=1,\dots,m$, and define the measures μ_j by

$$\mu_j(A)=\lambda_j^{-1}\,\mu(A\cap A_j),$$

where $j\in J$ and A is any Borel subset of K.

Clearly $\mu_j\in M_1^+(K)$, $\operatorname{Supp}(\mu_j)\subset W_j\cap K$ for $j\in J$, and $\mu=\sum_{j\in J}\lambda_j\mu_j$. Let $x_j=x_{\mu_j}$, and observe that $x_j\in K\cap W_j$ since μ_j is a positive normalized measure supported by the convex compact set $K\cap W_j$ for every $j\in J$.

If $x\in W_j\cap K, j\in J$, then

$$x-x_j=(x-x'_j)+(x'_j-x_j)\in W_0+W_0\subset W,$$

and so by (2.15)

$$|f_i(x)-f_i(x_j)|<\varepsilon, \qquad i=1,\dots,n.$$

Since μ_j is a positive normalized measure supported by W_j, we obtain

$$|\mu_j(f_i)-f_i(x_j)|<\varepsilon, \qquad i=1,\dots,n;\ j\in J. \tag{2.16}$$

Now define $\nu=\sum_{j\in J}\lambda_j\varepsilon_{x_j}$. Clearly $\nu\in M_1^+(K)$ and $x_\nu=x_\mu$. By (2.16) we obtain for $i=1,\dots,n$:

$$\begin{aligned}|\mu(f_i)-\nu(f_i)|&=\Big|\sum_{j\in J}\Big(\int_{A_j}f_i\,d\mu-\lambda_j f_i(x_j)\Big)\Big|\\&\leqq\sum_{j\in J}\lambda_j|\mu_j(f_i)-f_i(x_j)|\leqq\varepsilon.\end{aligned}$$

This completes the proof. □

Corollary I.2.4. *A point x in a compact convex set K is extreme iff ε_x is the only measure in $M_1^+(K)$ with barycenter x.*

* It is of interest to note that the barycenter formula (2.13) subsists for affine functions of the first Baire class, but is inexact for affine Baire functions of higher classes.

In the proof of these facts we shall apply the following notation for the *oscillation* of a (possibly extended) real valued function f *over a subset* Y of its domain X:

$$\mathrm{O}f(Y) = \sup_{x,y \in Y} |f(x) - f(y)|. \tag{2.17}$$

Also we shall introduce a notation for the *oscillation* of f *at a point* x of X, when X is a topological space:

$$\mathrm{O}f(x) = \inf\{\mathrm{O}f(U) \mid x \in U,\ U \text{ is open}\}. \tag{2.18}$$

Lemma I.2.5. *Let f be a bounded function on a compact convex set K, let μ be a probability measure on K, and let $\varepsilon > 0$. Assume that for every Borel subset B of K with $\mu(B) > 0$, there exists a Borel set $D \subset B$ such that $\mu(D) > 0$ and*

$$\mathrm{O}f(\mathrm{cl.conv}\, D) \leqq \varepsilon. \tag{2.19}$$

Then there exists a (finite or infinite) sequence $\{D_n\}$ of pairwise disjoint Borel subsets of K such that

$$\mathrm{O}f(\mathrm{cl.conv} D_n) \leqq \varepsilon, \qquad n = 1, 2, \ldots, \tag{2.20}$$

$$\sum_n \mu(D_n) = 1. \tag{2.21}$$

Proof. We first define a non-negative set function Λ on the Borel sets:

$$\Lambda(B) = \sup\{\mu(D) \mid D \in \mathscr{B},\, D \subset B,\, \mathrm{O}f(\mathrm{cl.conv}\, D) \leqq \varepsilon\}. \tag{2.22}$$

Using the hypothesis of the Lemma, we may construct inductively a sequence $\{D_n\}$ of pairwise disjoint Borel sets satisfying (2.20) and

$$\mu(D_n) > (1 - 2^{-n})\Lambda(K \setminus (D_1 \cup \cdots \cup D_{n-1})), \qquad n = 1, 2, \ldots, \tag{2.23}$$

and such that the sequence breaks off after its n-th term iff

$$\mu(D_1 \cup \cdots \cup D_n) = 1. \tag{2.24}$$

If the sequence $\{D_n\}$ is finite, then (2.21) is trivial. Otherwise, (2.23) gives

$$\Lambda(K \setminus (D_1 \cup \cdots \cup D_{n-1})) < 2\mu(D_n),$$

which implies

$$\lim_{n\to\infty} \Lambda(K\setminus(D_1 \cup \cdots \cup D_n)) = 0. \tag{2.25}$$

Now assume that $\sum_n \mu(D_n) < 1$. Then $B = K \setminus \bigcup_n D_n$ is a Borel set of positive measure. By hypothesis there is a Borel set $D \subset B$ such that $\mu(D) > 0$ and (2.19) is satisfied. By the definition of Λ

$$\Lambda(K\setminus(D_1 \cup \cdots \cup D_n)) \geqq \mu(D), \qquad n = 1, 2, \ldots$$

This contradicts (2.25), and the proof is complete. □

Theorem I.2.6. *(Choquet) If f is a real valued affine function of the first Baire class on a compact convex set K in a locally convex Hausdorff space E, then f is bounded, and*

$$f(x) = \int f \, d\mu,$$

where μ is any probability measure on K and x is the barycenter of μ.

Proof. 1. The set of discontinuity points of f is of first category in K (Osgood's Theorem, cf. e. g. [240]). By Baire's Theorem, f has a continuity point y in K. In particular, f is bounded in $K \cap U$ for some open neighbourhood U of y. By compactness, K is a bounded set, and so it can be absorbed by U. Hence there is a natural number m such that

$$K \subset y + m(U - y).$$

Since f is affine, it must be bounded on K.

2. We claim that the assumptions of Lemma I.2.5 are satisfied for the given f and μ, and for any $\varepsilon > 0$.

Let B be some Borel subset of K such that $\mu(B) > 0$, and consider the restriction $\nu = \mu | B$, defined by $\nu(A) = \mu(A \cap B)$ for all $A \in \mathscr{B}$. Let $C = \mathrm{Supp}(\nu)$ an $J = \mathrm{cl.\,conv.\,C}$. From now on we shall work within J, and we define $g = f | J$ and

$$Y = \left\{ x \in J \mid O g(x) \geqq \frac{\varepsilon}{3} \right\}. \tag{2.26}$$

It follows from the definition of the oscillation of a function at a point that Y is closed, and it follows from the affine nature of g that Y is convex. Again we may apply Osgood's Theorem to conclude that there is a continuity point of g in J. Hence $J \setminus Y \neq \emptyset$. Since Y is closed and convex, we shall even have $C \setminus Y \neq \emptyset$.

Let $z \in C \setminus Y$ and let U be a closed convex neighbourhood of z such that $U \cap Y = \emptyset$. By assumption $z \in C = \operatorname{Supp}(\nu)$, and so $\nu(U \cap K) \neq 0$. Writing $V = U \cap J$, we shall have $\nu(V) = \nu(U \cap K) \neq 0$, and $V \cap Y = \emptyset$. By (2.26)

$$O g(x) < \frac{\varepsilon}{3}, \quad \text{all } x \in V. \tag{2.27}$$

By boundedness, the range of f may be decomposed into a finite union of disjoint intervals I_j, each of length less than $\varepsilon/3$. This induces a decomposition of V into a finite union of disjoint convex Borel sets $V_j = V \cap g^{-1}(I_j)$ such that

$$O g(V_j) \leqq \frac{\varepsilon}{3}. \tag{2.28}$$

We claim that for every index j:

$$O g(\bar{V}_j) \leqq \varepsilon. \tag{2.29}$$

In fact for any two elements x, y of $\bar{V}_j$, we may choose neighbourhoods W_x, W_y of x, y in J such that $O g(W_x) < \varepsilon/3$, $O g(W_y) < \varepsilon/3$ (cf. (2.27)). Choosing $x' \in V_j \cap W_x$, $y' \in V_j \cap W_y$, and using (2.28) and the triangle inequality, we obtain the relation $|g(x) - g(y)| \leqq \varepsilon$, which proves (2.29).

Now $\sum_j \nu(V_j) = \nu(V) \neq 0$ and so there is at least one index k such that $\nu(V_k) \neq 0$. We define $D = V_k \cap B$, and obtain $\nu(D) = \nu(V_k) \neq 0$. Moreover it follows from the fact that $\bar{V}_k$ is convex, that cl. conv. $D \subset \bar{V}_k$. Hence by (2.29)

$$O f(\text{cl. conv. } D) = O g(\text{cl. conv. } D) \leqq \varepsilon,$$

and the requirements of Lemma I.2.5 is satisfied.

Let $\{D_n\}$ be a sequence of disjoint Borel subsets of K each of positive μ-measure, satisfying (2.20) and (2.21). Choose a natural number N such that

$$\lambda_0 = \sum_{n > N} \mu(D_n) \leqq \varepsilon, \tag{2.30}$$

and define $\lambda_n = \mu(D_n)$ for $n = 1, \ldots, N$. Define $\mu_n = \lambda_n^{-1} \mu | D_n$ for $n = 1, \ldots, N$; define $\mu_0 = \lambda_0^{-1} \mu | (K \setminus (D_1 \cup \cdots \cup D_N))$ if $\lambda_0 \neq 0$ and $\mu_0 = 0$ if $\lambda_0 = 0$.

Clearly we can express μ as a convex combination $\mu = \sum_{n=0}^{N} \lambda_n \mu_n$. Writing $x_n = x_{\mu_n}$, and using the fact that μ_n is concentrated on D_n, we obtain $x_n \in$ cl. conv. D_n for $n = 1, \ldots, N$.

By (2.20) this gives

$$|f(x_n) - \mu_n(f)| \leqq \varepsilon, \quad n = 1, \ldots, N. \tag{2.31}$$

Clearly $x = x_\mu = \sum_{n=0}^{N} \lambda_n x_n$, and so by (2.30), (2.31), and by the affine nature of f:

$$|f(x) - \mu(f)| = \left| \sum_{n=0}^{N} \lambda_n f(x_n) - \sum_{n=0}^{N} \lambda_n \mu_n(f) \right|$$
$$\leqq \sum_{n=0}^{N} \lambda_n |f(x_n) - \mu_n(f)|$$
$$\leqq \varepsilon(2\|f\| + 1).$$

This completes the proof since $\varepsilon > 0$ was arbitrary. □

This may be the appropriate place to introduce semicontinuous multivalued functions. Their use will simplify the proof of our next proposition, and they are indispensable in the study of continuous selections [291], [258], [261], [271].

A *multivalued function* Ψ from a set S to a set T is a map which assigns to every point $s \in S$ a subset Ψ_s of T. If S and T are topological spaces, then Ψ is *u.s.c.* if

$$U \quad \text{open in } T \Rightarrow \{s \in s \mid \Psi_s \subset U\} \quad \text{open,} \tag{2.32}$$

and Ψ is *l.s.c.* if

$$U \quad \text{open in } T \Rightarrow \{s \in S \mid \Psi_s \cap U \neq \emptyset\} \quad \text{open.} \tag{2.33}$$

Lemma I.2.7. *Let Ψ be a l.s.c. multivalued function from a topological space S into a topological space T and let φ be a continuous real valued function on $S \times T$. Then the function $f: S \to \mathbb{R}$ defined by*

$$f(s) = \inf\{\varphi(s,t) \mid t \in \Psi_s\}, \tag{2.34}$$

is u.s.c.

Proof. Let $\alpha \in \mathbb{R}$. To prove that $\{s \in S \mid f(s) < \alpha\}$ is open, we assume $f(s_0) < \alpha$ and we shall find a neighbourhood of s_0 in which this inequality prevails.

By definition there is a point $t_0 \in \Psi_{s_0}$ such that $\varphi(s_0, t_0) < \alpha$, and by continuity there are neighbourhoods V and U of s_0 and t_0, respectively, such that

$$(s,t) \in V \times U \to \varphi(s,t) < \alpha. \tag{2.35}$$

Now $t_0 \in \Psi_{s_0} \cap U$, and by (2.33) there exists a neighbourhood V' of s_0 such that $\Psi_s \cap U \neq \emptyset$ for $s \in V'$. Hence if $s \in V \cap V'$, then there exists a

$t \in \Psi_s \cap U$, and now $\varphi(s,t) < \alpha$ and $t \in \Psi_s$. Hence $s \in V \cap V'$ implies $f(s) < \alpha$, and the proof is complete. □

Remark. Lemma I.2.7 is just one out of four similar statements. A change of "inf" to "sup" in (2.34) induces a change from u.s.c. to l.s.c. in the conclusion. If $\{\varphi(s,t) | t \in T\}$ is an equicontinuous family in $C(S)$ and $\{\varphi(s,t) | s \in S\}$ is a family in $C(T)$, then the assumption that φ be u.s.c. in Lemma I.2.7 implies that $f(s)$ is l.s.c. Similarly, a change of „inf" to „sup" in (2.34) implies that $f(s)$ is u.s.c. Note that these statements apply to the special case where $T = \mathbb{R}$ and $\varphi(s,t) = t$. Now it is seen that a real valued function g on S is u.s.c. (l.s.c.) iff the associated multivalued function $s \rightsquigarrow \{\alpha \in \mathbb{R} | \alpha \leqslant g(s)\}$ is u.s.c. (l.s.c.).

Proposition I.2.8. *Let X be a compact Hausdorff space and define $D_\alpha(\mu) = \{x \in X \,|\, \mu(\{x\}) \geqslant \alpha\}$ for $\mu \in M_1^+(X)$ and $\alpha \in \mathbb{R}^+$. Then $\mu \rightsquigarrow \mu(D_\alpha(\mu))$ is a vaguely u.s.c. real valued function on $M_1^+(X)$ for every $\alpha > 0$.*

Proof. The proof proceeds in three steps.

1. First we show that $\mu \rightsquigarrow D_\alpha(\mu)$ is an u.s.c. multivalued function. This is the clue to the proof, and it is a precise formulation of the more intuitive statement that "the heavy atoms do not move much under a vague perturbation".

Let U be an open subset of X and assume $D_\alpha(\mu_0) \subset U$. By regularity we may assign to every $x \in U$ a neighbourhood V_x such that $\mu_0(\overline{V}_x) < \alpha$. Let $x_1, \ldots, x_n \in \complement U$ be chosen such that

$$\complement U \subset \bigcup_{i=1}^{n} V_{x_i},$$

and consider the set

$$W = \{\mu \in M_1^+(X) | \mu(\overline{V}_{x_i}) < \alpha, i = 1, \ldots, n\}. \tag{2.36}$$

This set is vaguely open in $M_1^+(X)$ because the function $\mu \rightsquigarrow \mu(\overline{V}_{x_i}) = \mu(\chi_{\overline{V}_{x_i}})$ is vaguely u.s.c. for $i = 1, \ldots, n$, since generally $\mu \rightsquigarrow \mu(f)$ is vaguely u.s.c. whenever f is u.s.c. (cf. (2.3)). Also $\mu_0 \in W$, and $D_\alpha(\mu) \subset U$ for every $\mu \in W$. Hence we have proved that $\mu \rightsquigarrow D_\alpha(\mu)$ is u.s.c.

2. Next we define a multivalued function Ψ of $M_1^+(X)$ into $C_{\mathbb{R}}(X)$ by

$$\Psi_\mu = \{g \in C_{\mathbb{R}}(X) | g \geqslant \chi_{D_\alpha(\mu)}\}, \tag{2.37}$$

and we shall prove that it is l.s.c.

Let U be an open subset of $C_{\mathbb{R}}(X)$ and assume $\Psi_{\mu_0} \cap U \neq \emptyset$, say

$$g_0 \geqslant \chi_{D_\alpha(\mu_0)}, \qquad g_0 \in U. \tag{2.38}$$

Let $\varepsilon > 0$ be chosen such that for $g \in C_{\mathbb{R}}(X)$:

$$\|g - g_0\|_\infty < \varepsilon \Rightarrow g \in U, \tag{2.39}$$

and define

$$V = \left\{ x \in X \,\middle|\, g_0(x) > 1 - \frac{\varepsilon}{2} \right\}. \tag{2.40}$$

Clearly V is an open set containing $D_\alpha(\mu_0)$. By the first part of the proof there is a vague neighbourhood W' of μ_0 such that $D_\alpha(\mu) \subset V$ for all $\mu \in W'$. Now let $\mu \in W'$ and $g = g_0 + \varepsilon/2$. Then $g \in U$ by virtue of (2.39), and it follows from (2.40) that

$$\chi_{D_\alpha(\mu)} \leqslant \chi_V \leqslant g\,.$$

Hence $g \in U \cap \Psi$, and we have proved that Ψ is l.s.c.

3. By the finiteness of $D_\alpha(\mu)$ the function $\chi_{D_\alpha(\mu)}$ is u.s.c. Hence we shall have

$$\begin{aligned} \mu(D_\alpha(\mu)) &= \inf \{\mu(g) \mid g \in C_{\mathbb{R}}(X),\, g \geqslant \chi_{D_\alpha(\mu)}\} \\ &= \inf \{\varphi(\mu, g) \mid g \in \Psi_\mu\}, \end{aligned}$$

where $\varphi(\mu, g) = \mu(g)$. Now the proposition follows by application of the second part of the proof and Lemma I.2.7. □

Remark. The compactness is not essential in Proposition I.2.8. To treat the general case, one should make use of the fact that for every probability Radon measure μ and every $\varepsilon > 0$ there exists a compact set C such that $\mu(\complement C) < \varepsilon$.

Corollary I.2.9. *Let X be a metrizable compact Hausdorff space and let $\varphi: M_1^+(X) \to \mathbb{R}$ be defined by $\varphi(\mu) = \mu_d(X)$, where μ_d denotes the discrete (purely atomic) component of μ. Then φ is of second Baire class with respect to the vaque topology of $M_1^+(X)$.*

Proof. Defining $\varphi_n(\mu) = \mu(D_{1/n}(\mu))$ we shall have $\varphi_n \nearrow \varphi$, and the statement follows from Proposition I.2.8. and the well-known fact that a semicontinuous function on a metrizable compact Hausdorff space is of first Baire class. □

Example I.2.10. *(Choquet) Let $K = M_1^+([0, 1])$ endowed with the vague topology, and let $\varphi: K \to \mathbb{R}$ be defined by $\varphi(\mu) = \mu_d([0, 1])$ where μ_d is the discrete part of μ. Then φ is a bounded Baire function of second class for which the barycenter formula is invalid.*

Proof. Clearly φ is bounded and affine. By the separability of $C([0,1])$, the unit ball of $M([0,1])$ is w^*-metrizable (cf. e.g. [140]). Hence K is metrizable, and so φ is of second Baire class by virtue of Corollary I.2.9.

It is well known that $\psi: t \rightsquigarrow \varepsilon_t$ is a homeomorphism of $[0,1]$ onto $\partial_e K$, and in particular that $\partial_e K$ is closed. Let $\psi\lambda$ be the direct image of Lebesgue measure λ by ψ, i.e. $[\psi\lambda](E) = \lambda(\psi^{-1}(E))$ for every Borel subset E of K, or equivalently

$$\int g\, d(\psi\lambda) = \int g \circ \psi\, d\lambda, \quad \text{all } g \in C(K). \tag{2.41}$$

Every w^*-continuous linear functional F on $M([0,1])$ is the evaluation at some $f \in C_{\mathbb{R}}([0,1])$. Hence $F(\psi(t)) = [\psi(t)](f) = f(t)$ for all $t \in [0,1]$, and it follows from (2.41) that

$$\int F\, d(\psi\lambda) = \int f\, d\lambda = \lambda(f) = F(\lambda).$$

Hence λ is the barycenter of $\psi\lambda$.

The bounded Baire function φ is identically one on the set $\partial_e K = \{\varepsilon_x \mid x \in [0,1]\}$ which supports $\psi\lambda$. Hence we shall have

$$\int \varphi\, d(\psi\lambda) = 1, \qquad \varphi(\lambda) = 0, \tag{2.42}$$

and the proof is complete. □

Notes. In Proposition I.2.1. the hypothesis $f(T) \subset K$ where K is weakly compact, can be omitted in various special cases, e.g. if E is complete or quasicomplete (every closed and bounded subset is complete), and also if E is a Banach space endowed with the weak topology [140]. Hence every *weakly* continuous function f from a compact Hausdorff space T into a Banach space E is weakly μ-integrable for every $\mu \in M_{\mathbb{R}}(T)$.

Corollary I.2.4. was announced by H. Bauer [37].

In the article [104] G. Choquet raised the problem on the validity of the barycenter formula for affine Borel functions. In a post-script to the same article he proved Theorem I.2.6 and established Example I.2.9.

§ 3. Comparison of Measures on a Compact Convex Set

The space $N(K)$ of affine dependences on a compact convex set K induces an equivalence relation on $M_{\mathbb{R}}(K)$. Specifically $\mu \sim \nu$ iff $\mu - \nu \in N(K)$, or equivalently:

$$\mu(K) = \nu(K), \qquad \int x\, d\mu(x) = \int x\, d\nu(x). \tag{3.1}$$

The subspace $N(K)$ is the annihilator in $M_{\mathbb{R}}(K)$ of $A(K)$, and so every $q \in A(K)^*$ is the common restriction to $A(K)$ of all measures in an equivalence class, which we shall denote by $M_q(K)$. Thus $\mu \in M_q(K)$ iff

$$q(a) = \int a(x)\, d\mu(x), \quad \text{all } a \in A(K). \tag{3.2}$$

If q is the evaluation at a point $x \in K$, i.e. if $q(a) = a(x)$ for $a \in A(K)$, then we shall write $M_x(K)$ in the place of $M_q(K)$. Thus $\mu \in M_x(K)$ iff

$$a(x) = \int a(y)\, d\mu(y), \quad \text{all } a \in A(K), \tag{3.3}$$

and in this case we shall say that the measure μ *represents* the point x.

We shall use the symbol $M_x^+(K)$ to denote the set of all positive measures μ which represent the point $x \in K$.

Evaluating at the constant 1, we obtain $\|\mu\| = \mu(K) = 1$ for $\mu \in M_x^+(K)$. Hence $M_x^+(K) \subset M_1^+(K)$, and it follows that $M_x^+(K)$ is a *vaguely compact*, convex set for every $x \in K$. Note that if $\mu \in M_1^+(K)$, then the statement $\mu \in M_x^+(K)$ is equivalent to $\mu \sim \varepsilon_x$, which in turn is equivalent to $x_\mu = x$.

Let $P(K)^*$ be the dual cone in $M_{\mathbb{R}}(K)$ of the cone $P(K)$ of continuous convex functions on a compact convex set K, i.e. $\mu \in P(K)^*$ iff $\mu(f) \geqq 0$ for all $f \in P(K)$. The order relation defined by $P(K)^*$ will be denoted by the symbol $\prec$. Thus

$$\mu \prec \nu \Leftrightarrow \mu(f) \leqq \nu(f) \quad \text{all } f \in P(K). \tag{3.4}$$

Here the anti-symmetry (i.e. $\mu \prec \nu \prec \mu \Rightarrow \mu = \nu$) follows by the density of $P(K) - P(K)$ in $\mathscr{C}_{\mathbb{R}}(K)$ (Prop. I.1.1).

Applying (3.4) to a and $-a$, we obtain

$$\mu \prec \nu \Rightarrow \mu(a) = \nu(a) \quad \text{all } a \in A(K). \tag{3.5}$$

Hence *μ and ν are comparable only if* $\mu \sim \nu$.

Observe that a simple limit argument based on Cor. I.1.3 and (2.3) transforms (3.4) into

$$\mu \prec \nu \Leftrightarrow \mu(f) \leqq \nu(f) \quad \text{all } f \in Q(K). \tag{3.6}$$

If $\mu, \nu \in M_1^+(K)$ and $\mu \prec \nu$, then one may think of μ and ν as two distributions of positive unit mass with the same barycenter (since $\mu \sim \nu$) and such that ν assumes the greater value at every continuous convex function. Loosely one may say that "the mass of ν is removed farther away from the common barycenter of μ and ν, and comes closer to the extreme boundary of K".

We have a simple, but useful, characterization of the relation $\mu \prec \nu$ for positive measures:

Proposition I.3.1. *Let μ, ν be positive measures on a compact convex set K. Now $\mu \prec \nu$ iff*

$$\mu(\check{f}) \leqq \nu(f), \tag{3.7}$$

or equivalently

$$\nu(f) \leqq \mu(\hat{f}), \tag{3.8}$$

for all $f \in C_{\mathbb{R}}(K)$. In fact, in (3.7) *one may allow f to be l.s.c. and in* (3.8) *one may allow f to be u.s.c.*

Proof. 1. If $\mu \prec \nu$ and $f \in C_{\mathbb{R}}(K)$, then $\check{f} \in Q(K)$, and so by (3.6)

$$\mu(\check{f}) \leqq \nu(\check{f}) \leqq \nu(f).$$

Conversely, if we specialize (3.7) to an $f \in P(K)$, we obtain

$$\mu(f) = \mu(\check{f}) \leqq \nu(f).$$

The equivalence of (3.7) and (3.8) follows by the substitution $f \rightsquigarrow -f$.

2. To prove the last statements, we shall need to know that if $\{f_\alpha\}$ is a net from $C_{\mathbb{R}}(K)$ and $f_\alpha \uparrow f$, then $\check{f}_\alpha \uparrow \check{f}$. In fact, assume $x_0 \in K$ and $\beta < \check{f}(x_0)$. By Prop. I.1.2, there is an $a \in A(K)$ such that

$$a < \check{f}, \qquad a(x_0) - \beta = \delta > 0.$$

Now $\{a \wedge f_\alpha\}$ is an increasing net from $C_{\mathbb{R}}(K)$ which converges pointwise to $a \in C_{\mathbb{R}}(K)$. By Dini's Lemma the convergence is uniform, and so there is an index α_0 such that $f_{\alpha_0} \geqq a - \delta$. It follows that $\check{f}_{\alpha_0} \geqq a - \delta$, and so in particular

$$\check{f}_{\alpha_0}(x_0) \geqq a(x_0) - \delta = \beta.$$

Thus we have proved that $\sup_\alpha \check{f}_\alpha \geqq \check{f}$, which is the non-trivial part of the statement $\check{f}_\alpha \uparrow \check{f}$.

Now we assume (3.7) valid for continuous functions, and we consider a l.s.c. function f on K. Let $\{f_\alpha\}$ be some net from $C_{\mathbb{R}}(K)$ such that $f_\gamma \uparrow f$. By (2.3) and by the statement just proved:

$$\mu(\check{f}) = \sup_\alpha \mu(\check{f}_\alpha) \leqq \sup_\alpha \nu(f_\alpha) = \nu(f).$$

Again the dual statement follows by the substitution $f \rightsquigarrow -f$. □

Proposition I.3.2. *(Cartier, Fell, Meyer) Let μ, ν be positive measures on a compact convex set K. Then $\mu \prec \nu$ iff for every decomposition $\mu = \sum_{i=1}^{n} \mu_i$ of μ into n positive components μ_i there exists a decomposition*

$\nu = \sum_{i=1}^{n} \nu_i$ *of* ν *into* n *positive components* ν_i *such that* $\nu_i \sim \mu_i$ *for* $i = 1, \ldots, n$; *in this case one can even take* $\mu_i \prec \nu_i$ *for* $i = 1, \ldots, n$.

Proof. 1. Assume first $\mu \prec \nu$, and $\mu = \sum_{i=1}^{n} \mu_i$ where $\mu_i \geqq 0$ for $i = 1, \ldots, n$. Define a map $\Phi : C_{\mathbb{R}}(K)^n \to \mathbb{R}$ by

$$\Phi(\tilde{f}) = \sum_{i=1}^{n} \mu_i(\hat{f}_i), \tag{3.9}$$

where $\tilde{f} = (f_1, \ldots, f_n)$. It follows by (1.7), (1.8) that Φ is sub-linear.

Let F be the vector-subspace of $C_{\mathbb{R}}(K)^n$ consisting of all $\tilde{f} = (f, \ldots, f)$, and define a linear functional Ψ_0 on F by $\Psi_0(\tilde{f}) = \nu(f)$ for $f \in C_{\mathbb{R}}(K)$. It follows by (3.8) that

$$\Psi_0(\tilde{f}) = \nu(f) \leqq \mu(\hat{f}) = \sum_{i=1}^{n} \mu_i(\hat{f}) = \Phi(\tilde{f})$$

for all $\tilde{f} = (f, \ldots, f)$ where $f \in C_{\mathbb{R}}(K)$.

By the Hahn-Banach Theorem there exists a linear functional Ψ on $C_{\mathbb{R}}(K)^n$ which extends Ψ_0 and remains bounded by Φ. Now Ψ is bounded in the norm $\|\tilde{f}\| = \max_{1 \leqq i \leqq n} \|f_i\|$ where $\tilde{f} = (f_1, \ldots, f_n)$; in fact $\|\Psi\| \leqq \|\mu\|$.

The bounded linear functional Ψ on $C_{\mathbb{R}}(K)^n$ can be expressed as follows:

$$\Psi(\tilde{f}) = \sum_{i=1}^{n} \nu_i(f_i), \quad \tilde{f} = (f_1, \ldots, f_n) \tag{3.10}$$

where $\nu_i \in C_{\mathbb{R}}(K)^* = M_{\mathbb{R}}(K)$. Observe that $\nu_i \geqq 0$.

Appling (3.10) to $\tilde{f} = (f_1, \ldots, f_n)$ where $f_i = 0$ for $i \neq k$, $f_k = f$, we obtain

$$\nu_k(f) = \Psi(\tilde{f}) \leqq \Phi(\tilde{f}) = \mu_k(\hat{f}) \tag{3.11}$$

for every $f \in C_{\mathbb{R}}(K)$, $k = 1, \ldots, n$.

By Prop. I.3.1 $\mu_k \prec \nu_k$, and so in particular $\mu_k \sim \nu_k$ for $k = 1, \ldots, n$.

2. We assume the condition of the proposition, and we shall prove that $\mu \prec \nu$, i.e. we shall prove that $\mu(f) \leqq \nu(f)$ for an arbitrary $f \in P(K)$.

Let $\varepsilon > 0$, and let $K \subset \bigcup_{i=1}^{n} G_i$ where G_i is closed and convex and

$$x, y \in G_i \Rightarrow |f(x) - f(y)| \leqq \varepsilon, \qquad i = 1, \ldots, n. \tag{3.12}$$

We write $A_i = G_i \setminus \bigcup_{j<i} G_j$ and define $\mu_i = \mu | A_i$ for $i = 1, \ldots, n$. Now $\mu = \sum_{i=1}^{n} \mu_1$, and by assumption there is a decomposition $\nu = \sum_{i=1}^{n} \nu_i$ where $\nu_i \geqq 0$ and $\mu_i \sim \nu_i$. We may assume that all $\mu_i \neq 0$.

Let x_i be the common barycenter of μ_i and ν_i for $i=1,\ldots,n$. Then $x_i \in$ cl. conv. $A_i \subset G_i$, and so by (3.12)

$$\mu_i(f) \leqq \mu_i(K)(f(x_i)+\varepsilon), \qquad i=1,\ldots,n. \tag{3.13}$$

Applying the formula (2.11) we obtain

$$\mu(f)=\sum_{i=1}^{n}\mu_i(f) \leqq \sum_{i=1}^{n}\nu_i(K)(f(x_i)+\varepsilon)$$

$$\leqq \sum_{i=1}^{n}(\nu_i(f)+\varepsilon\cdot\nu_i(K))=\nu(f)+\varepsilon\cdot\nu(K).$$

Since $\varepsilon>0$ was arbitrary, $\mu(f)\leqq\nu(f)$, and the proof is complete. □

The next Corollary is a mere reformulation of the above Proposition, and the following Corollary is a simple specialization of the first.

Corollary I.3.3. *Let μ, ν be positive, normalized measures on a compact convex set K. Then $\mu \prec \nu$ iff for every convex combination $\mu=\sum_{i=1}^{n}\lambda_i\mu_i$ where $\mu_i \in M_1^+(K)$ there exists a corresponding convex combination $\nu=\sum_{i=1}^{n}\lambda_i\nu_i$ where $\nu_i \in M_1^+(K)$ and $\nu_i \sim \mu_i$ for $i=1,\ldots,n$, in this case one can even take $\mu_i \prec \nu_i$ for $i=1,\ldots,n$.*

Corollary I.3.4. *Let $\mu,\nu \in M_1^+(K)$, where K is a compact convex set. If μ is a simple measure, say $\mu=\sum_{i=1}^{n}\lambda_i\varepsilon_{x_i}$ where $\lambda_i \in \mathbb{R}^+$ and $\sum_{i=1}^{n}\lambda_i=1$, then $\mu \prec \nu$ iff ν can be written as $\nu=\sum_{i=1}^{n}\lambda_i\nu_i$ where $\nu_i \in M_1^+(K)$ and $x_{\nu_i}=x_i$, or what is equivalent $\nu_i \in M_{x_i}^+(K)$, for $i=1,\ldots,n$.*

If $\mu,\nu \in M_1^+(K)$, $\mu \prec \nu$ and μ is a *simple* measure, then one may visualize ν as the result of an "explosion" of each of the point-masses of μ. Briefly one may state this by saying that ν is a "dilation" of μ. A definition of "dilation" for general measures in $M_1^+(K)$ will be given in the starred section at the end of the present paragraph.

It follows from Corollary I.3.4 above that *a measure $\nu \in M_1^+(K)$ satisfies a relation $\varepsilon_x \prec \nu$ iff $\nu \in M_x^+(K)$*. This can of course also be easily verified by a direct argument.

Our next proposition is the main tool in the development leading to Choquet's Representation Theorem.

Proposition I.3.5. *Let K be a compact convex set, $f \in C_{\mathbb{R}}(K)$ and $\mu \in M_1^+(K)$. Then there exists a measure $\nu \in M_1^+(K)$ such that $\mu \prec \nu$ and $\nu(f) = \mu \hat{f})$.*

Proof. By virtue of (1.7), (1.8) the functional $\Phi: g \rightsquigarrow \mu(\hat{g})$ is sublinear on $C_{\mathbb{R}}(K)$. Over the 1-dimensional subspace of $C_{\mathbb{R}}(K)$ spanned by f, we define a linear functional ν_0 by writing

$$\nu_0(\alpha f) = \alpha \mu(\hat{f}), \quad \text{all } \alpha \in \mathbb{R}. \tag{3.14}$$

We claim that ν_0 is majorized by Φ.

For $\alpha \geqq 0$ we have by (1.8):

$$\nu_0(\alpha f) = \alpha \mu(\hat{f}) = \mu(\widehat{\alpha f}) = \Phi(\alpha f).$$

For $\alpha < 0$ we write $\alpha = -\beta$, and we first observe that the sub-additivity (1.7) applied to the sum of βf and $-\beta f$ yields $-\mu(\widehat{\beta f}) \leqq \mu(\widehat{-\beta f})$. Hence for $\alpha < 0$:

$$\nu_0(\alpha f) = -\beta \mu(\hat{f}) = -\mu(\widehat{\beta f}) \leqq \mu(\widehat{-\beta f}) = \mu(\widehat{\alpha f}) = \Phi(\alpha f).$$

By the Hahn-Banach Theorem there exists a linear functional ν on $C_{\mathbb{R}}(K)$ which extends ν_0 and remains majorized by Φ. Hence

$$\nu(f) = \mu(\hat{f}); \ \nu(g) \leqq \mu(\hat{g}) \quad \text{all } g \in C_{\mathbb{R}}(K). \tag{3.15}$$

If $g \in C_{\mathbb{R}}(K)$ and $g \leqq 0$, then (by (1.5)) $\hat{g} \leqq 0$, and so

$$\nu(g) \leqq \mu(\hat{g}) \leqq 0.$$

Hence ν is a positive linear functional on $C_{\mathbb{R}}(K)$, or in other words $\nu \in M^+(K)$.

By (3.15) and by Proposition I.3.1, $\mu \prec \nu$. In particular $\mu \sim \nu$, and it follows that ν is normalized since μ is normalized. This completes the proof. □

Corollary I.3.6. *Let K be a compact convex set, $f \in C_{\mathbb{R}}(K)$ and $x \in K$. Then:*

$$\begin{aligned} \hat{f}(x) &= \sup\{\nu(f) \mid \nu \in M_x^+(K)\} \\ &= \sup\left\{ \sum_{i=1}^{n} \lambda_i f(x_i) \,\middle|\, x = \sum_{i=1}^{n} \lambda_i x_i \ \text{(convex sum)} \right\}, \end{aligned} \tag{3.16}$$

and the first supremum is attained for some $\nu \in M_x^+(K)$.

The dual statements are valid for $\check{f}$.

Proof. The first equality of (3.16) is equivalent to

$$\varepsilon_x(\hat{f}) = \sup\{\nu(f) \mid \nu \in M_1^+(K),\ \varepsilon_x \prec \nu\},$$

which is valid by Proposition I.3.1 and Proposition I.3.5: in fact the supremum value is attained for some ν.

The last equality of (3.16) follows from the first by Proposition I.2.3. ☐

Remark. The first inequality of (3.16) subsists for an u.s.c. function f. (See Prop. 5.6 of [192].)

*Let K be a compact convex set, and let $\mu, \nu \in M_1^+(K)$. We shall say that ν is a *dilation* of μ if there exists a map $x \rightsquigarrow \lambda_x$ defined μ-almost everywhere on K with values in $M_1^+(K)$ satisfying

$$x \rightsquigarrow \lambda_x(f) \quad \text{is measurable for every } f \in C_{\mathbb{R}}(K), \tag{3.17}$$

$$\nu(f) = \int \lambda_x(f)\, d\mu(x) \quad \text{for every } f \in C_{\mathbb{R}}(K), \tag{3.18}$$

$$\lambda_x \in M_x^+(K) \quad \text{for } \mu\text{-almost all } x \text{ in } K. \tag{3.19}$$

Corollary I.3.4 states that if $\mu, \nu \in M_1^+(K)$ and μ is a *simple* measure, then ν is a dilation of μ iff $\mu \prec \nu$. We shall prove the corresponding theorem for a general μ on a *metrizable* K, by application of the following known theorem [89] on *disintegration of measure:*

Let C and K be metrizable compact (Hausdorff) spaces, let $\Phi: C \to K$ be a continuous surjection, and let $\vartheta \in M^+(C)$. Then there exists a map $x \rightsquigarrow \lambda_x$ from K into $M_1^+(C)$ such that

$$x \rightsquigarrow \lambda_x(f) \quad \text{is measurable for every } f \in C_{\mathbb{R}}(C). \tag{3.20}$$

$$\vartheta(f) = \int \lambda_x(f)\, d(\Phi\vartheta) \quad \text{for every } f \in C_{\mathbb{R}}(C). \tag{3.21}$$

$$\operatorname{Supp}(\lambda_x) \subset \Phi^{-1}(x) \quad \text{for every } x \in K. \tag{3.22}$$

In this connection we recall that the *direct image* $\Phi\vartheta$ of ϑ by Φ is characterized by the equality

$$(\Phi\vartheta)(A) = \vartheta(\Phi^{-1}(A)), \tag{3.23}$$

valid for all Borel subsets A of K; or equivalently by the equality

$$\int f\, d(\Phi\vartheta) = \int (f \circ \Phi)\, d\vartheta, \tag{3.24}$$

valid for all continuous functions f on K.

Lemma I.3.7. *Let K be a compact convex set, and define two subsets of $M_1^+(K) \times M_1^+(K)$ as follows*

$$R = \{(\mu, \nu) \mid \mu \prec \nu\}, \qquad S = \{(\varepsilon_x, \nu) \mid \nu \in M_x^+(K)\}. \tag{3.25}$$

These sets are, both compact (in the product of the vague topology by itself), R is convex, and $\partial_e R \subset S \subset R$.

Proof. The non-trivial part of the proof is the verification that $\partial_e R \subset S$. Assume $(\mu, \nu) \in \partial_e R \setminus S$. Then μ is not a one-point measure, and so it can be written as a convex combination $\mu = \lambda \mu_1 + (1-\lambda)\mu_2$ where $0<\lambda<1$; $\mu_1, \mu_2 \in M_1^+(K)$, and $\mu_1 \neq \mu_2$. By Corollary I.3.3 one may decompose ν into a convex combination $\nu = \lambda \nu_1 + (1-\lambda)\nu_2$ where $\nu_1, \nu_2 \in M_1^+(K)$ and $\mu_1 \prec \nu_1, \mu_2 \prec \nu_2$. Hence $(\mu_1, \nu_1), (\mu_2, \nu_2) \in R$, and the decomposition

$$(\mu, \nu) = \lambda(\mu_1, \nu_1) + (1-\lambda)(\mu_2, \nu_2)$$

contradicts the assumption that (μ, ν) be extreme in R. □

Theorem I.3.8. *(Cartier) Let K be a metrizable compact convex set and $\mu, \nu \in M_1^+(K)$. The measure ν is a dilation of the measure μ iff $\mu \prec \nu$.*

Proof. 1. Assume first that $\mu \prec \nu$. By Lemma I.3.7, there exists a measure $\vartheta_0 \in M_1^+(S)$ which represents the point (μ, ν). (Integral form of the Krein-Milman Theorem, cf. e.g. [41] or [331]). For any two functions $f, g \in C_{\mathbb{R}}(K)$, the functional $(\varphi, \psi) \rightsquigarrow \varphi(f) + \psi(g)$ on R belongs to $A(R)$, and so

$$\mu(f) + \nu(g) = \int_S [\varphi(f) + \psi(g)]\, d\vartheta_0(\varphi, \psi). \tag{3.26}$$

There is a natural isomorphism between S and the set

$$C = \{(x, \psi) \mid x \in K, \psi \in M_x^+(K)\}.$$

Denoting the image of ϑ_0 under this isomorphism by ϑ, we shall have

$$\mu(f) + \nu(g) = \int_S [f(x) + \psi(g)]\, d\vartheta(x, \psi), \tag{3.27}$$

for all $f, g \in C_{\mathbb{R}}(K)$.

Now we consider the continuous map $\Phi: C \to K$ defined by $\Phi(x, \psi) = x$, and we claim that the direct image of ϑ by Φ is the given measure μ. In fact we may apply (3.27) with an arbitrary $f \in C_{\mathbb{R}}(K)$ and $g = 0$, to obtain

$$(\Phi\vartheta)(f) = \int_C (f \circ \Phi)\, d\vartheta = \int_C f(x)\, d\vartheta(x, \psi) = \mu(f).$$

Hence $\Phi\vartheta = \mu$ as claimed.

Let $\{\lambda_x^0\}_{x \in K}$ be a family of measures in $M_1^+(C)$ which disintegrates ϑ with respect to Φ. For an arbitrary $g \in C_{\mathbb{R}}(K)$ we apply (3.21) to the function $F_g \in C_{\mathbb{R}}(C)$ defined by $F_g(x, \psi) = \psi(g)$, and we obtain:

$$\vartheta(F_g) = \int_K \lambda_x^0(F_g)\, d\mu(x). \tag{3.28}$$

For every $x \in K$ we define $\lambda_x : g \rightsquigarrow \lambda_x^0(F_g)$. It is easily verified that λ_x is a positive and normalized linear functional on $C_{\mathbb{R}}(K)$. Hence $\lambda_x \in M_1^+(K)$ for all $x \in K$. Also we observe that $x \rightsquigarrow \lambda_x(g) = \lambda_x^0(F_g)$ is measurable by virtue of (3.20). Hence the family $\{\lambda_x\}_{x \in K}$ satisfies the requirement (3.17) in the definition of a dilation.

At this point we return to the formula (3.27), and now we apply it with an arbitrary $g \in C_{\mathbb{R}}(K)$ and $f = 0$. By the definitions of F_g and λ_x, and by (3.28) we obtain

$$\begin{aligned} \nu(g) &= \int_S \psi(g)\, d\vartheta(x,\psi) = \vartheta(F_g) \\ &= \int_K \lambda_x^0(F_g)\, d\mu(x) = \int_K \lambda_x(g)\, d\mu(x). \end{aligned}$$

Hence $x \rightsquigarrow \lambda_x$ satisfies the crucial requirement (3.18) in the definition of a dilation.

It remains to be verified that for every $x \in K$ one has $\lambda_x \in M_x^+(K)$, or equivalently that

$$\lambda_x(a) = a(x), \quad \text{all } a \in A(x). \tag{3.31}$$

By (3.22), λ_x^0 is supported by $\Phi^{-1}(x) \subset C$, and by the definitions of Φ and C

$$\Phi^{-1}(x) = \{(x,\psi) \mid \psi \in M_x^+(K)\}, \quad \text{all } x \in K.$$

Now observe that a *simple* measure $\pi \in M_1^+(\Phi^{-1}(x))$ is of the form

$$\pi = \sum_{i=1}^{n} \alpha_i \varepsilon_{(x,\psi_i)} \quad \text{(convex sum)}, \tag{3.32}$$

with $\psi_i \in M_x^+(K)$ for $i = 1, \ldots, n$.

It follows by the definition of F_a that

$$\pi(F_a) = \sum_{i=1}^{n} \alpha_i \psi_i(a) = a(x).$$

By an elementary theorem (cf. e.g. [87]) λ_x^0 is vague limit of measures π of the form (3.32). Hence

$$\lambda_x^0(F_a) = a(x), \quad \text{all } a \in A(K).$$

By the definition of λ_x this gives (3.31), and the first part of the proof is complete.

2. Assume that ν is a dilation of μ and that $x \rightsquigarrow \lambda_x$ is a mapping of K into $M_1^+(K)$ satisfying (3.17)–(3.19). If $g \in P(K)$, then by (3.19) and by Proposition I.2.2

$$g(x) \leqq \lambda_x(g),$$

for μ-almost all $x \in K$.

By (3.18) this gives

$$\nu(g)=\int \lambda_x(g)\,d\mu(x)\geqq \int g(x)\,d\mu(x).$$

Hence $\nu(g)\geqslant\mu(g)$ for every $g\in P(K)$, and so $\mu\prec\nu$. The proof is complete. □

Notes. Order relations like the relation $\mu\prec\nu$ have appeared in various papers by various authors. The definition (3.4) was given by Choquet in 1960 [101], but the idea of proving the Choquet Theorem by means of an ordering of measures, appeared also in Bishop-de Leeuw's paper of 1959 [63]. They used a slightly different ordering in which the cone $P(K)$ was replaced by the (smaller) set of all squares a^2 with $a\in A(K;E)$.

The conclusion of Proposition I.3.2 was taken as the definition of a "strong ordering" of measures (denoted by $\mu<<\nu$) by Loomis in 1962 [282], and this relation was used by him in a very natural proof of the Choquet Uniqueness Theorem. (Cf. a subsequent chapter of the present book.) The mutual equivalence of "$\mu\prec\nu$" and "$\mu<<\nu$" was established by Cartier, Fell and Meyer in 1964 [96].

Proposition I.3.5 is the main tool in the construction of the "maximal measures" (or "boundary measures") occuring in the integral theorems of Choquet and Bishop- de Leeuw, to which we shall return in the next paragraph. This proposition is essentially an application of the analytic form of the Hahn-Banach Theorem, and it is not clear to whom it should most rightly be contributed. However, it occurs quite explicitely in the works of Choquet-Meyer [111] and of Bonsall [81], both from 1963.

The concept of "dilation of measure" has roots going quite far back. In their well known treatise on inequalities from 1934, Hardy, Littlewood and Polya proved the conclusion of our Theorem I.3.8 for simple measures on a one-dimensional space (cf. [208]). It was proved for arbitrary probability measures on a compact convex set in $\mathbb{R}^1$ (a closed interval) in 1950 by Blackwell [66], and it was proved for a simple probability measure on a compact convex set in $\mathbb{R}^n$ in 1951 by Sherman [369]. The general statement of Theorem I.3.8 was conjectured by Choquet in 1960 [101], and the first proof was published in the article by Cartier, Fell, Meyer in 1964 [96]. In his book on Probability and Potentials Meyer ascribes this theorem to Cartier [289] and in [96] it is mentioned that Mokobodzki independently had obtained the same result (unpublished). In 1965, Strassen generalized Cartier's Theorem, using a rather different approach [377]. His method of proof is of considerable interest in itself.

§ 4. Choquet's Theorem

We shall first prove a simple, but useful, characterization of extreme points:

Proposition I.4.1. *(Hervé) Let x be a point in a compact convex set K. Then the following statements are equivalent:*

(i) $x \in \partial_e K$.
(ii) $f(x) = \hat{f}(x)$ *for all* $f \in C_{\mathbb{R}}(K)$.
(iii) $f(x) = \hat{f}(x)$ *for every u.s.c. real function* $f: K \to [-\infty, \infty[$.

One may replace $\hat{f}$ *by* $\check{f}$ *in* (ii), *and also in* (iii) *if u.s.c. is changed to l.s.c.*

Proof. (i) ⇔ (ii). Assume first $x \in \partial_e K$ and consider an $f \in C_{\mathbb{R}}(K)$. By Proposition I.3.5, there is a measure $f \in M_x^+(K)$ such that $\nu(f) = \hat{f}(x)$. By Corollary I.2.4, $M_x^+(K) = \{\varepsilon_x\}$, and so $f(x) = \hat{f}(x)$.

Conversely, assume (ii) and consider an $f \in C_{\mathbb{R}}(K)$. Passing from f to $-f$, we obtain $f(x) = \hat{f}(x) = \check{f}(x)$. Let $\nu \in M_x^+(K)$ be arbitrary. By Proposition I.3.1

$$f(x) = \check{f}(x) \leqslant \nu(f) \leqq \hat{f}(x) = f(x),$$

and so $\nu = \varepsilon_x$. Hence $M_x^+(K) = \{\varepsilon_x\}$, and by Corollary I.2.4 x is extreme.

(ii) ⇔ (iii). Assume (ii) and let f be bounded and u.s.c. There is a descending net $\{f_\alpha\}$ from $C_{\mathbb{R}}(K)$ which converges pointwise to f, and $\inf_\alpha f_\alpha$ is seen to be u.s.c. and concave. Hence it majorizes $\hat{f}$, and so

$$f(x) = \inf_\alpha f_\alpha(x) = \inf_\alpha \hat{f}_\alpha(x) \geqslant \hat{f}(x).$$

This completes then non-trivial part of the proof. □

Corollary I.4.2. *Let f be a bounded continuous real function defined on a subset X of a compact convex set K, and assume that* $\partial_e K \subset X$. *Then:*

$$f(x) = \hat{f}(x) = \check{f}(x), \quad \text{all } x \in \partial_e K. \tag{4.1}$$

Proof. Let $f(x) \geqslant \alpha \in \mathbb{R}$ for all $x \in X$, and extend f to an u.s.c. function f' on K as follows:

$$f'(x) = \begin{cases} \limsup\limits_{y \in X \; y \to x} f(y), & \text{if } x \in \bar{X}, \\ \alpha, & \text{if } x \in K \setminus \bar{X}. \end{cases}$$

Clearly every continuous concave function on K which majorizes f on X, also majorizes f' on the whole convex set K. Hence $\hat{f} = \hat{f}'$, and the first equality of (4.1) follows from the above Proposition I.4.1. The verification of the second equality is similar. □

To every continuous real function f on a compact convex set K we associate a *boundary set* B_f defined by

$$B_f = \{x \in K \mid \hat{f}(x) = f(x)\}. \tag{4.2}$$

Clearly

$$B_f = \bigcap_{n=1}^{\infty} \left\{ x \,\middle|\, \hat{f}(x) - f(x) < \frac{1}{n} \right\}, \tag{4.3}$$

and it follows that *the boundary set of a continuous function is a* G_δ*-set.*

Note also that by Proposition I.4.1:

$$\partial_e K = \bigcap_{f \in C_{\mathbb{R}}(K)} B_f. \tag{4.4}$$

A point x of a compact convex set K is said to be a point of *strict convexity* for a real valued function f on K if

$$f(x) < \lambda f(y) + (1-\lambda) f(z),$$

for every proper convex combination $x = \lambda y + (1-\lambda) z$ (i.e. $y, z \in K$, $y \neq z$, $0 < \lambda < 1$).

Note that *the extreme points of* K *are automatically points of strict convexity*. (The above requirement is vacuously satisfied.)

Note also that *if* x *is a non-extreme point of strict convexity for an* $f \in C_{\mathbb{R}}(K)$, *then* $x \notin B_f$. In fact, if $x \in B_f$ then we should have

$$f(x) = \hat{f}(x) \geqslant \lambda \hat{f}(y) + (1-\lambda) \hat{f}(z) \geqslant \lambda f(y) + (1-\lambda) f(z),$$

for any convex combination $x = \lambda y + (1-\lambda) z$.

A real valued function f on a compact convex set K is said to be *strictly convex* if every point in K is a point of strict convexity for f.

By the above remarks:

$$\partial_e K = B_f \tag{4.5}$$

for every strictly convex function $f \in P(K)$ (if any).

Theorem I.4.3. *(Hervé) A compact convex set* K *admits a strictly convex, continuous real function iff* K *is metrizable.*

Proof. 1. Assume first that K is metrizable. By a known theorem [140], $C_{\mathbb{R}}(K)$ is separable, and by Proposition I.1.1 it has a dense sequence $\{g_n^1 - g_n^2\}$ where $g_n^1, g_n^2 \in P(K)$ for $n=1,2,\ldots,$ We rename the functions g_n^1, g_n^2 and arrange them in a single sequence $\{f_n\}$.

Now we claim that the continuous function

$$f = \sum_{n=1}^{\infty} 2^{-n} \|f_n\|^{-1} f_n$$

is strictly convex.

If this were not the case, then there would be a proper convex combination $x = \lambda_y + (1-\lambda)z$ on K such that

$$f(x) = \lambda f(y) + (1-\lambda) f(z). \tag{4.6}$$

By the definition of f, the corresponding equality would hold with any f_n in the place of f, and hence it would hold for any function in $C_{\mathbb{R}}(K)$, which is absurd. (Consider the square of a continuous affine function separating y and z.)

2. Assume next that K admits a strictly convex continuous real function f, and define a continuous function $\omega \geqslant 0$ on $K \times K$ by

$$\omega(x,y) = \frac{1}{2}[f(x) + f(y)] - f\left(\frac{x+y}{2}\right). \tag{4.7}$$

By the metrization theorem for uniform spaces (cf. e.g. [239]), it suffices to prove that the sequence $\{W_n\}$ of sets

$$W_n = \left\{(x,y) \,\middle|\, x,y \in K,\, \omega(x,y) < \frac{1}{n}\right\} \tag{4.8}$$

generates the uniformity of K.

Recall that the uniformity of K has a base consisting of the sets

$$V_U = \{(x,y) \mid x,y \in K,\, x-y \in U\}, \tag{4.9}$$

where U is an open balanced neighbourhood of the origin in the locally convex Hausdorff space E in which K is embedded. Note that V_U is open in $K \times K$.

The function f is uniformly continuous on K. Hence for every natural number n there is an open balanced neighbourhood U of the origin of E such that

$$x,y \in K,\, x-y \in U \Rightarrow |f(x) - f(y)| < \frac{1}{n}.$$

It follows that for $x,y \in K,\, x-y \in U$:

$$\omega(x,y) \leqslant \frac{1}{2}\left|f(x) - f\left(\frac{x+y}{2}\right)\right| + \frac{1}{2}\left|f(y) - f\left(\frac{x+y}{2}\right)\right| < \frac{1}{n}.$$

Thus we have proved $V_U \subset W_n$.

To prove the reverse relation, we consider an arbitrary V_U of the form (4.9). It follows by strict convexity that $\omega(x,y)\neq 0$ if $x-y\neq 0$, and in particular if $(x,y)\notin V_U$. Hence

$$\omega(K\times K\setminus V_U)\subset \mathbb{R}^+\setminus\{0\},$$

and by compactness there is a natural number n such that

$$\omega(K\times K\setminus V_U)\subset\left[\frac{1}{n},\infty\right[.$$

This means that $W_n\subset V_U$, and the proof is complete. □

From Theorem I.4.3 and formula (4.5) we obtain the following:

Corollary I.4.4. *If K is a metrizable compact convex set, then there is an $f\in P(K)$ such that $\partial_e K=B_f$. In particular, $\partial_e K$ will be a G_δ-set.*

Remark. One may give a short direct proof of the last assertion of the above Corollary I.4.4 by considering the direct images of the sets $\{(x,y)\,|\,\mathrm{dist}(x,y)\geqslant n^{-1}\}$ under the map $(x,y)\rightsquigarrow\frac{1}{2}(x+y)$. Their union is equal to $K\setminus\partial_e K$, which is thereby an F_σ-set.

Proposition I.4.5. *(Mokobodzki) Let μ be a positive measure on a compact convex set K. Then the following statements are equivalent:*

(i) *μ is a maximal element of $M^+(K)$ with respect to the ordering "$\prec$".*

(ii) *$\mu(\hat{f})=\mu(f)$ for all $f\in C_{\mathbb{R}}(K)$.*

(iii) *$\mu(\hat{f})=\mu(f)$ for all $f\leqslant P(K)$.*

One may replace $\hat{f}$ by $\check{f}$ in (ii), *and also in* (iii) *if $P(K)$ is changed to $-P(K)$.*

Proof. (i) ⇒ (ii) Assume (i) and let $f\in C_{\mathbb{R}}(K)$ be arbitrary. By Proposition I.3.5, there exists a positive measure ν such that $\mu\prec\nu$ and $\nu(f)=\mu(\hat{f})$. By maximality $\mu=\nu$, and so $\mu(f)=\mu(\hat{f})$.

(ii) ⇒ (iii) Evident.

(iii) ⇒ (i) Assume (iii) and let ν be any positive measure such that $\mu\prec\nu$. If $f\in P(K)$, then $f=\check{f}$, and it follows by Proposition I.3.1 that

$$\mu(f)=\mu(\check{f})\leqslant\nu(f)\leqslant\mu(\hat{f})=\mu(f).$$

Hence μ and ν coincide on $P(K)$, and by Proposition I.1.1 they coincide on $C_{\mathbb{R}}(K)$. Hence $\nu=\mu$, and it follows that μ is maximal in $M^+(K)$. □

Remark 1. It is easily verified that the equation $\mu(f)=\mu(\hat{f})$ subsists also for a maximal positive measure μ an u.s.c. function f. In fact, for every $\varepsilon>0$ there exists a $g\in C_{\mathbb{R}}(K)$ such that $g\geqq f$ and

$$\mu(f)+\varepsilon\geqq\mu(g)=\mu(\hat{g})\geqq\mu(f).$$

Note, however, that the equation may fail if f is l.s.c, as can be seen by simple two-dimensional examples.

2. It is worth noting that Proposition I.4.1 is a special case of the above result, obtained by taking $\mu=\varepsilon_x$ for some $x\in K$.

A measure μ on a compact convex set K is said to be a *boundary measure* if $|\mu|$ satisfies the three (equivalent) requirements of Proposition I.4.5.

Clearly the statement (ii) of Proposition I.4.5 is equivalent to $\mu(K\backslash B_f)=0$. Thus a measure μ on K is a boundary measure iff

$$|\mu|(K\backslash B_f)=0, \quad \text{all } f\in C_{\mathbb{R}}(K); \tag{4.10}$$

i.e. if it vanishes off every boundary set B_f *where* $f\in C_R(K)$.

It follows from Corollary I.4.4 that *a measure* μ *on a metrizable compact convex set* K *is a boundary measure iff it vanishes off the* G_δ*-set* $\partial_e K$, i.e. if

$$|\mu|(K\backslash\partial_e K)=0. \tag{4.11}$$

Note that $\partial_e K$ need not be measurable in the non-metrizable case, so it is not meaningful to ask if (4.11) prevails in general. In the sequel we shall return to the question if a boundary measure on a general compact convex set K is in some sense "concentrated" on $\partial_e K$.

At present we note that *a simple measure* $\mu=\sum_{i=1}^{n}\alpha_i\varepsilon_{x_i}$ where $\alpha_i\neq 0$ *for* $i=1,\dots,n$, *is a boundary measure on a general compact convex set* K *iff* $x_i\in\partial_e K$ *for* $i=1,\dots,n$. This is in fact an easy consequence of Proposition I.4.1.

We also note the following *necessary* condition for μ being a boundary measure:

Proposition I.4.6. *If* μ *is a boundary measure on a compact convex set* K, *then* $\operatorname{Supp}(\mu)\subset\overline{\partial_e K}$.

Proof. Let C be a compact subset of K such that $C\cap\overline{\partial_e K}=\emptyset$, and let $f\in\mathscr{C}_{\mathbb{R}}(K)$ take values in $[0,1]$ such that $f(\overline{\partial_e K})=0, f(C)=1$. Now since $\check{f}=0$,

$$|\mu|(C)\leqq|\mu|(f)=\mu|(\check{f})=0.$$

It follows (by regularity) that $|\mu|(K\backslash\overline{\partial_e K})=0$, and the proof is complete. ∎

Lemma I.4.7. *For every positive measure μ on a compact convex set K there exists a positive boundary measure ν such that $\mu \prec \nu$.*

Proof. We define $M_\mu^+ = \{\nu \in M^+(K) \mid \mu \prec \nu\}$, and we observe that $\|\nu\| = \nu(1) = \mu(1) = \|\mu\|$ for all $\nu \in M_\mu^+(K)$. We claim that M_μ^+ is inductive in the ordering "$\prec$".

To this end we consider a generalized sequence $\{\nu_\alpha\}_{\alpha\in A}$ from M_μ^+, and we assume that the index set A is linearly ordered and that $\{\nu_\alpha\}_{\alpha\in A}$ is *ascending*, in that $\alpha < \beta \Rightarrow \nu_\alpha \prec \nu_\beta$. By vague compactness $\{\nu_\alpha\}_{\alpha\in A}$ has vague accumulation point ν_0 in the set of all $\nu \in M^+(K)$ with $\|\nu\| \leqslant \|\mu\|$. Let $\alpha \in A$ and $f \in P(K)$. For every $\varepsilon > 0$ there is an index $\beta > \alpha$ such that ν_β is in the vague (f, ε)-neighbourhood ν_0, that is

$$|\nu_\beta(f) - \nu_0(f)| < \varepsilon. \tag{4.12}$$

By assumption $\nu_\alpha \prec \nu_\beta$, and so $\nu_\alpha(f) \leqslant \nu_\beta(f)$. Hence by (4.12)

$$\nu_\alpha(f) - \nu_0(f) < \varepsilon. \tag{4.13}$$

Since $\varepsilon > 0$ was arbitrary, this gives $\nu_\alpha(f) \leqslant \nu_0(f)$. Since $f \in P(K)$ was arbitrary, we shall have $\mu \prec \nu_\alpha \prec \nu_0$. Thus ν_0 is an upper bound of $\{\nu_\alpha\}_{\alpha\in A}$ in $M_\mu^+ K)$.

By Zorn's Lemma, $M_\mu^+(K)$ has a maximal element, and the proof is complete. □

Applying Lemma I.4.7 with $\mu = \varepsilon_x$, we immediately obtain the following:

Theorem I.4.8. *(Choquet-Bishop-de Leeuw) Every point x in a compact convex set can be represented by a positive and normalized boundary measure.*

Corollary I.4.9. *(Choquet) Every point x in a metrizable compact convex set K can be represented by a positive and normalized measure vanishing off the G_δ-set $\partial_e K$.*

We return to the investigation of boundary measures on general compact convex sets. The main tool is a maximum principle for superior limits of sequences from $Q(K)$.

Proposition I.4.10. *Let K be a compact convex set and let $f = \limsup_n f_n$ where $\{f_n\}$ is an upper bounded sequence from $Q(K)$. If $f(x) \leqslant \alpha$ for all $x \in \partial_e K$, then $f(x) \leqslant \alpha$ for all $x \in K$. The dual statement is valid for inferior limits of lower bounded sequences from $-Q(K)$.*

Proof. Let $x\in K$ be arbitrary. By Proposition I.1.2, there is a sequence $\{a_n\}$ from $A(K;E)$ such that

$$a_n \leqslant f_n,\ f_n(x) < a_n(x) + \frac{1}{n}, \tag{4.14}$$

for $n=1,2,\ldots$.

The continuous affine mapping $\Phi: y \mapsto \{a_n(y)\}$ of E into $\mathbb{R}^{\aleph_0}$ maps K onto a *metrizable* compact convex set K'.

For every natural number n we denote the n-th (canonical) projection in $\mathbb{R}^{\aleph_0}$ by p_n. Then p_n is a continuous linear functional on $\mathbb{R}^{\aleph_0}$ and $p_n \circ \Phi = a_n$ for $n=1,2,\ldots$.

We claim that

$$\limsup_n p_n(y') \leqslant \alpha, \tag{4.15}$$

for every $y' \in \partial_e K'$. In fact, $K \cap \Phi^{-1}(y')$ is a closed face of K; it contains an extreme point y (Krein-Milman), and $y \in \partial_e K$ (since $K \cap \Phi^{-1}(y')$ is a face). Now $y' = \Phi(y)$, and so by assumption

$$\limsup_n p_n(y') = \limsup_n a_n(y) \leqslant \alpha.$$

By Corollary I.4.9 there is a measure $\mu \in M_1^+(K')$ which represents the point $x' = \Phi(x)$, and for which $\mu(K' \backslash \partial_e K') = 0$.

The sequence $\{p_n\}$ is upper bounded on $K' = \Phi(K)$, since $\{a_n\}$ is upper bounded on K. Hence we can use Fatou's Lemma for superior limits, and by (4.14) and (4.15) we shall have:

$$\begin{aligned} f(x) &= \limsup_n a_n(x) = \limsup_n p_n(x') \\ &= \limsup_n \mu(p_n) \leqslant \mu\left(\limsup_n p_n\right) \leqslant \alpha. \end{aligned}$$

The proof is complete. □

Corollary I.4.11. *(Rainwater) If $\{a_n\}$ is a bounded sequence from the space $A(K)$ over a compact convex set, and $\lim_n a_n(x) = 0$ for all $x \in \partial_e K$, then $\lim_n a_n(x) = 0$ for all $x \in K$; consequently 0 is in the uniformly closed convex hull of $\{a_1, a_2, \ldots\}$.*

Corollary I.4.12. *(Bishop-de Leeuw) If μ is a boundary measure on a compact convex set K, then $|\mu|(C) = 0$ for every closed G_δ-set C disjoint from $\partial_e K$.*

Proof. By Urysohn's Lemma there is a bounded sequence $\{f_n\}$ from $\mathscr{C}_{\mathbb{R}}(K)^+$ such that $f_n(x) = 1$ for $x \in C$ and for all $n = 1, 2, \ldots$ and such that $\lim_n f_n(x)=0$ for $x\notin C$. In particular $|\mu|(C)\leqq|\mu|(f_n)$ for $n=1,2,\ldots$.

The hypotheses of Proposition I.4.10 is satisfied with $\check{f}_n$ in the place of f_n and with $\alpha=0$. Using this fact and the definition of boundary measure, we obtain by Fatou's Lemma

$$\begin{aligned} 0 &\leqslant |\mu|(C) \leqslant \limsup_n |\mu|(f_n) \\ &= \limsup_n |\mu|(\check{f}_n) \\ &\leqslant |\mu|\big(\limsup_n \check{f}_n\big) \leqq 0. \end{aligned}$$

This completes the proof. □

Remark. It follows by formula (2.2) and the subsequent remarks that the set C of Corollary I.4.12 can be replaced by any Baire set disjoint from $\partial_e K$. Observe also that it follows from the proof of this corollary that the conclusion subsist for any closed set C (and hence by regularity for any Borel set) contained in a G_δ-set disjoint from $\partial_e K$. However, it need not subsist for a Borel set disjoint from $\partial_e K$. (Cf. counter-example in the "starred" section at the end of this paragraph.)

In the rest of this section we shall have to take into account some of the measure theoretic distinctions mentioned at the beginning of § 2. Recall that the word *measure* is used primarily to denote a Radon measure (i.e. a continuous linear functional on the space of continuous functions), while the terms *Baire measure* and *Borel measure* are used to denote countably additive set-functions (on the σ-fields $\mathscr{B}_0$ and $\mathscr{B}$, respectively).

Corollary I.4.13. *If K is a compact convex set and $\mu\in M_1^+(K)$ is a boundary measure, then there exists a positive, countably additive set function $\bar{\mu}$ on the induced σ-field $\partial_e K\cap\mathscr{B}_0$ such that*

$$\mu(f) = \int_{\partial_e K} f\,d\bar{\mu}, \quad \text{all } f\in\mathscr{C}_{\mathbb{R}}(K), \tag{4.16}$$

and

$$\bar{\mu}(\partial_e K)=1. \tag{4.17}$$

Proof. Let μ_0 be the Baire measure on K corresponding to the given Radon measure μ, and write

$$\bar{\mu}(\partial_e K\cap B)=\mu_0(B), \quad \text{all } B\in\mathscr{B}_0. \tag{4.18}$$

We claim that $\bar{\mu}$ is well defined.

In fact let $B_1, B_2 \in \mathscr{B}_0$ and $B_1 \cap \partial_e K = B_2 \cap \partial_e K$. If $B_1 \triangle B_2$ denotes the symmetric difference, and $\varepsilon > 0$ is arbitrary, then there is a compact G_δ-set C such that $C \subset B_1 \triangle B_2$ and

$$\mu_0(B_1 \triangle B_2) < \mu(C) + \varepsilon.$$

(In this connection cf. (2.2) and the comments to that formula.)

By the assumption $B_1 \cap \partial_e K = B_2 \cap \partial_e K$, the set C is disjoint from $\partial_e K$. It follows from Corollary I.4.12 that $\mu_0(C) = 0$. Since ε was arbitrary, $\mu_0(B_1 \triangle B_2) = 0$, and so $\mu_0(B_1) = \mu_0(B_2)$. Hence $\bar{\mu}$ is well defined.

It is easily verified that $\bar{\mu}$ is positive and countably additive. Also

$$\bar{\mu}(\partial_e K) = \bar{\mu}(\partial_e K \cap K) = \mu_0(K) = 1.$$

To verify (4.16) we consider for a moment the σ-field $\mathscr{F}_0$ on K generated by $\mathscr{B}_0$ and the set $\partial_e K$. It consists of the sets

$$(B_1 \cap \partial_e K) \cup (B_2 \backslash \partial_e K); \; B_1, B_2 \in \mathscr{B}_0. \tag{4.19}$$

By the above argument the formula

$$\rho[(B_1 \cap \partial_e K) \cup (B_2 \backslash \partial_e K)] = \mu_0(B_1) \tag{4.20}$$

defines a countably additive set function ρ on $\mathscr{F}_0$ which reduces to μ_0 on $\mathscr{B}_0$ (take $B_1 = B_2$), and which vanishes off $\partial_e K$ and has the restriction $\bar{\mu}$ to $\partial_e K$. Since $\mathscr{B}_0 \subset \mathscr{F}_0$, every $f \in \mathscr{C}_{\mathbb{R}}(K)$ will be $\mathscr{F}_0$-measurable, and

$$\mu(f) = \int_K f \, d\mu_0 = \int_K f \, d\rho = \int_{\partial_e K} f \, d\bar{\mu}.$$

This completes the proof. □

Applying Theorem I.4.8, we obtain the following:

Theorem I.4.14. *(Bishop-de Leeuw) For every point x in a compact convex set K there exists a positive countably additive set function $\bar{\mu}$ on the σ-field $\partial_e K \cap \mathscr{B}_0$ such that*

$$a(x) = \int_{\partial_e K} a \, d\bar{\mu}, \quad \text{all } a \in A(K), \tag{4.21}$$

and

$$\bar{\mu}(\partial_e K) = 1. \tag{4.22}$$

Remark. The conclusion of Theorem I.4.14 was originally formulated in terms of the measure ρ which was introduced provisionally in the proof of Corollary I.4.13 (formula (4.20)). Note, however, that it is impossible to replace $\mathscr{B}_0$ by $\mathscr{B}$ in Corollary I.4.13 and in Theorem I.4.14, and that, in cases where the latter set is a Borel set, the (possibly non-regular) extension of μ_0 from $\mathscr{B}_0$ to $\mathscr{F}_0$ need not coincide with the regular Borel extension $\tilde{\mu}_0$ on the set. $\partial_e K$. (Cf. the "starred" section below.)

* We shall give an example of a compact convex set K where $\partial_e K$ is a (non-Baire) Borel set and $\tilde{\mu}_0(\partial_e K) \neq 1$, $\tilde{\mu}_0$ being the regular Borel measure corresponding to some boundary (Radon-) measure $\mu \in M_1^+(K)$.

For a subset A of a compact convex set K, the *σ-convex hull of A* is the set

$$\sigma(A) = \left\{ \sum_{i=1}^{\infty} \lambda_i x_i \,\middle|\, \sum_{i=1}^{\infty} \lambda_i = 1;\ \lambda_i \geqslant 0, x_i \in A, i = 1, 2, \ldots \right\}. \tag{4.23}$$

The summation to infinity in (4.23) indicates convergence in the given topology of K, which by compactness will coincide with the weak topology on K induced from the space E^*, dual to the locally convex Hausdorff space E in which K is embedded. Clearly, $\sigma(A)$ consists of all barycenters of *discrete* (purely atomic) measures $\mu \in M_1^+ K)$ "concentrated" on A (i.e. with all its atoms in A).

Proposition I.4.15. *There exist compact convex sets K where $\partial_e K$ is a Borel set in K, which are discrete in relative topology, and where $\sigma(\partial_e K) \neq K$.*

Proof. Consider a set $Y = \bigcup \{Y_\alpha | \alpha \in [0,1]\}$ where the sets Y_α are mutually disjoint, and each of them consists of just three points, say $Y_\alpha = \{r_\alpha, s_\alpha, t_\alpha\}$ for $\alpha \in [0,1]$.

We topologize Y in such a way that r_α, t_α are isolated points for all $\alpha \in [0,1]$, and such that each s_α has a neighbourhood base consisting of the sets

$$\{s_\alpha\} \cup \bigcup \{Y_\alpha | 0 < |\alpha - \beta| < \varepsilon\}, \qquad \varepsilon > 0.$$

The relative topology on each Y_α is seen to be discrete, and the mapping $s: \alpha \to s_\alpha$ is seen to be a homeomorphic embedding of $[0,1]$ into Y. It is also easy to verify that Y is compact. (The topology of Y is a special case of Bishop-de Leeuw's "porcupine topology" [63]).

Let N be the vaguely closed linear subspace of $M(Y)$ generated by the measures

$$\varepsilon_{s_\alpha} - \tfrac{1}{2}\,(\varepsilon_{r_\alpha} + \varepsilon_{t_\alpha}), \qquad \alpha \in [0,1] \tag{4.24}$$

Let $K = \varphi(M_1^+(Y))$ where φ is the canonical mapping of $M(Y)$ onto $E = M(Y)/N$, endowed with the quotient topology obtained from the vague topology of $M(Y)$.

We claim that (the restriction of) φ is a homeomorphism of the discrete (in the relative topology) set

$$Z = \{\varepsilon_{r_\alpha} | \alpha \in [0,1]\} \cup \{\varepsilon_{t_\alpha} | \alpha \in [0,1]\} \tag{4.25}$$

onto $\partial_e K$.

To prove this claim, we consider a point $z \in Z$, say $z = \varepsilon_{r_\alpha}$ for some $\alpha \in [0,1]$, and we define a function $f: Y \to \mathbb{R}$ by

$$f(y) = \begin{cases} 1 & \text{for } y = r_\alpha, \\ -1 & \text{for } y = t_\alpha, \\ 0 & \text{elsewhere}. \end{cases} \tag{4.26}$$

It is easily verified that $f \in \mathscr{C}_{\mathbb{R}}(Y)$ and that $\nu(f) = 0$ for all $\nu \in N$. Hence there is a well defined continuous linear functional F on E such that

$$F(\varphi(\mu)) = \mu(f), \quad \text{all } \mu \in M(Y). \tag{4.27}$$

It follows by the definitions that $F(\varphi(z)) = 1$ and $F(\varphi(z')) \leqslant 0$ for $z' \in Z \setminus \{z\}$. Hence $\varphi(z) \neq \varphi(z')$ for $z' \in Z \setminus \{Z\}$, and so $\varphi | Z$ is $1-1$. Also it follows that $\{\varphi(z') | z' \in Z, F(\varphi(z')) > 0\}$ is a neighbourhood of $\varphi(z)$ in $\varphi(Z)$ which does not contain any point $\varphi(z')$ with $z' \in Z \setminus \{z\}$. Hence $\varphi(z)$ is an isolated point of $\varphi(Z)$, and so $\varphi | Z$ is a homeomorphism of the set Z onto the set $\varphi(Z)$, both being discrete in relative topology.

To prove that every point of $\varphi(Z)$, say $\varphi(\varepsilon_{r_\alpha})$ for $\alpha \in [0,1]$, is extreme in K, we consider a convex combination

$$\varphi(\varepsilon_{r_\alpha}) = \gamma\, \varphi(\mu) + (1-\gamma)\, \varphi(\nu), \qquad 0 < \gamma < 1 \tag{4.28}$$

where $\mu, \nu \in M_1^+(Y)$.

Applying the functional F of (4.27) we obtain

$$1 = \gamma\, \mu(f) + (1-\gamma)\, \nu(f).$$

By the definition (4.26) of f, this is possible only if $\mu = \nu = \varepsilon_{r_\alpha}$. Hence the convex combination (4.28) must be improper, and so $\varphi(\varepsilon_{r_\alpha}) \in \partial_e K$.

To prove that φ maps Z onto $\partial_e K$, we consider an arbitrary point $u \in \partial_e K$. The set $\varphi^{-1}(u) \cap M_1^+(Y)$ is a closed face of $M_1^+(Y)$. It contains an extreme point, which must also be extreme in the whole set $M_1^+(Y)$: hence of the form ε_y for some $y \in Y$. We claim that $y \neq s_\alpha$ for all $\alpha \in [0,1]$. In fact if $y = s_\alpha$ for some $\alpha \in [0,1]$, then by the definition (4.24) of N

$$u = \varphi(\varepsilon_{s_\alpha}) = \tfrac{1}{2}\varphi(\varepsilon_{r_\alpha}) + \tfrac{1}{2}\varphi(\varepsilon_{t_\alpha}),$$

and by the preceding part of the proof $\varphi(\varepsilon_{r_\alpha}) \neq \varphi(\varepsilon_{t_\alpha})$. This gives a contradiction since $u \in \partial_e K$. Hence $\varepsilon_y \in Z$ and $\varphi(\varepsilon_y) = u$, which completes the verification that $\varphi(Z) = \partial_e K$.

The above argument also shows that φ maps the two sets Z and $S=\{\varepsilon_{s_\alpha}|\alpha\in[0,1]\}$ onto disjoint subsets of K. Hence

$$\partial_e K=\varphi(Z)=\varphi(S\cup Z)\setminus\varphi(S).$$

Clearly $S\cup Z$ and S are compact subsets of $M_1^+(Y)$ (in the vague topology). Hence $\varphi(S\cup Z)$ and $\varphi(S)$ are compact in E, and it follows that $\partial_e K$ is a Borel set.

It remains to be proved that $\sigma(\partial_e K)\neq K$. To this end we consider the Lebesgue measure λ on $[0,1]$ and its direct image $\tilde{\lambda}$ on Y under the mapping $s:[0,1]\to Y$. Clearly $\tilde{\lambda}\in M_1^+(Y)$. We claim that $\varphi(\tilde{\lambda})\notin\sigma(\partial_e K)$.

To prove this claim, we assume the contrary, that is we assume that

$$\varphi(\tilde{\lambda})=\sum_{i=1}^{\infty}\lambda_i\varphi(z_i), \tag{4.29}$$

where $\sum_{i=1}^{\infty}\lambda_i=1$ and $\lambda_i\geqslant 0, z_i\in Z$ say $z_i=\varepsilon_{y_i}$ where $y_i=r_{\alpha_i}$ or $y_i=t_{\alpha_i}$ for $i=1,2,\ldots$.

Let m be a natural number such that

$$\sum_{i>m}\lambda_i<\tfrac{1}{2} \tag{4.30}$$

and let g be a continuous and real valued function on $[0,1]$ such that

$$0\leqslant g\leqslant 1,\qquad g(\alpha_1)=\cdots=g(\alpha_m)=0,\qquad \lambda(g)>\tfrac{1}{2}. \tag{4.31}$$

Now we define a function $h:Y\to\mathbb{R}$ by

$$h(y)=g(\alpha),\quad \text{all } y\in Y_\alpha \tag{4.32}$$

where α runs through $[0,1]$.

Clearly h is continuous, $0\leqslant h\leqslant 1$ and $\tilde{\lambda}(h)=\lambda(g)>\frac{1}{2}$. Observe also that $\nu(h)=0$ for all $\nu\in N$. Hence there is a well defined continuous linear functional H on E such that

$$H(\varphi(\mu))=\mu(h),\quad \text{all } \mu\in M(Y). \tag{4.33}$$

Now it follows by the explicit form (4.29) of $\varphi(\tilde{\lambda})$ and from (4.30)–(4.33) that

$$H(\varphi(\tilde{\lambda}))=\sum_{i>m}\lambda_i g(\alpha_i)<\tfrac{1}{2}\,.$$

On the other hand $H(\varphi(\tilde{\lambda}))=\tilde{\lambda}(h)>\frac{1}{2}$. This contradiction completes the proof. □

Proposition I.4.16. *A compact convex set K with the properties stated in Proposition* I.4.15, *will admit a boundary (Radon-) measure* $\mu \in M_1^+(K)$ *such that* $\tilde{\mu}_0(\partial_e K) \neq 1$, $\tilde{\mu}_0$ *being the corresponding regular Borel measure.*

Proof. Let $\mu \in M_1^+(K)$ be a boundary measure representing a point $x \notin \sigma(\partial_e K)$.

By regularity and the fact that $\partial_e K$ is discrete,

$$\tilde{\mu}_0(\partial_e K) = \sup\{\tilde{\mu}_0(C) | C \subset \partial_e K, C \text{ finite}\}.$$

Hence $\tilde{\mu}_0(\partial_e K) = 1$ would imply that μ was a discrete positive normalized measure concentrated on $\partial_e K$. This is impossible since μ represents $x \notin \sigma(\partial_e K)$. □

Remark. In the above proof one can even obtain $\tilde{\mu}_0(\partial_e K) = 0$ by passing to the continuous (non-atomic) part of the original measure. (The property of being a boundary measure is preserved.)

Notes. Proposition I.4.1 and Theorem I.4.3 were both proved by Hervé in 1961 [209], but the former occurred implicitely already in Kadison's memoir of 1951 [224].

Proposition I.4.5 was proved in Mokobodzki's paper [297] from 1962.

Theorem I.4.8 gives the most general form of the boundary integral theorem. It was stated and proved by Bishop and de Leeuw in 1959 [63]. The application of Zorn's Lemma used in the proof of Lemma I.4.7, is due to them. Corollary I.4.9 is Choquet's original theorem, which was stated at a Bourbaki Seminar already in 1956 [100].

Proposition I.4.10 is due to P. A. Meyer, and it appeared first in seminar notes from Paris dated 1961/62 [288].

Corollary I.4.11 was stated by Rainwater in 1963 [341] and proved by means of Theorem I.4.14.

The Corollaries I.4.12 and I.4.13 and also Theorem I.4.14 are due to Bishop and de Leeuw [63]. A measure with the property stated in the conclusion of Corollary I.4.12 is sometimes said to be "pseudo carried" by the extreme boundary. In Choquet-Meyer's paper from 1963 there is an example, ascribed to Mokobodzki, of a measure which is pseudo carried by the extreme boundary without being a boundary measure [111]. This example pertains also to the question of uniqueness of representing measures, and we shall return to it in a subsequent chapter. In the same paper of Choquet-Meyer the conclusion of Proposition I.4.12 is improved to yield $|\mu|(C) = 0$ for every $\mathscr{K}$-analytic (or "$\mathscr{K}$-Suslin") set C disjoint from $\partial_e K$ [111].

The idea of the example given in the "starred" section at the end of § 4, is due to Bishop and de Leeuw [63]. We have preferred to present it in a somewhat different form to make explicit the geometric content, as stated in Proposition I.4.15.

§ 5. Abstract Boundaries Defined by Cones of Functions

Fundamental concepts of the preceding paragraphs, like Choquet's ordering $\prec$ and the notion of a representing boundary measure, were all defined by means of the cone $P(K)$ of continuous convex functions on K. In the present paragraph we shall generalize to more general cones of functions defined on compact spaces which are not a priori endowed with any affine structure. Here we shall start with cones of u.s.c. functions in order to admit certain important applications, in particular to subharmonic functions.

To connect the abstract theory with the previous results, we first prove the following:

Proposition I.5.1. *(Mokobodzki) Let $f: K \to [-\infty, \infty[$ be an u.s.c. function on a compact convex subset K of a locally convex Hausdorff space E. For every $x \in K$ we shall have*

$$f(x) = \inf\{g(x) \mid g > f, g \in P(K)\}. \tag{5.1}$$

If f is convex, then there is a descending net $\{q_\alpha\}$ from $P(K)$ such that $q_\alpha \searrow f$.

Proof.
1. Let $\sup_{x \in K} f(x) = \alpha < \infty$, and let $x_0 \in K$. If $f(x_0) = \alpha$, then (5.1) is trivial. Assume therefore $f(x_0) < \alpha$, and let β be an arbitrary number such that $f(x_0) < \beta < \alpha$. By upper semi-continuity there is a closed symmetric and convex neighbourhood V of 0 in E such that $f(x) < \beta$ for all $x \in K \cap (x_0 + V)$. Let u be the Minkowski functional of V, and define

$$g(x) = \beta + (\alpha - \beta) u(x - x_0), \qquad x \in K.$$

Now it is seen that $g > f$ and $g(x_0) = \beta$. Clearly $g \in P(K)$, and the first part of the proof is complete.

2. We assume that f is convex, and we shall prove that for any two functions $g_1, g_2 \in P(K)$ such that $g_1, g_2 > f$, there exists a $g \in P(K)$ such that $g_1, g_2 > g > f$.

We define $M = \{(x,\lambda) | x \in K, \lambda \leqslant f(x)\}$ and $N_i = \{(x,\lambda) | x \in K, g_i(x) \leqslant \lambda \leqslant \Lambda\}$ where $g_i(x) \leqslant \Lambda$ for all $x \in K$ and $i = 1,2$. Now M, N_1, N_2 are closed subsets of $E \times \mathbb{R}$, and N_1, N_2 are also convex and compact. It follows that the set $Q = \operatorname{conv}(N_1 \cup N_2)$ is compact, and by the convexity of f we shall have $Q \cap M = \emptyset$. By an elementary theorem on topological vector spaces (or even on uniform spaces) there exist an open neighbourhood W of 0 in $E \times \mathbb{R}$ such that $(Q+W) \cap (M+W) = \emptyset$.

Let $W = U \times]-\varepsilon, \varepsilon[$ where U is a convex open neighbourhood of 0 in E and $\varepsilon > 0$, and define a real valued function g on $K+U$ by

$$g(x) = \inf\{\lambda \in \mathbb{R} | (x,\lambda) \in Q + W\}.$$

Now g is seen to be convex, and $f < g < g_1, g_2$. Also g is continuous since g is an upper bounded convex function defined on the *open* convex set $K+U$ (Cf. e.g. [85]).

Corollary I.5.2. *Two non-negative measures μ, ν on a compact convex set K satisfy the requirement $\mu \prec \nu$ iff $\mu(f) \leqslant \nu(f)$ for every u.s.c. convex function $f: K \to [-\infty, \infty[$.*

In the rest of this paragraph we shall study cones S of functions defined over an arbitrary, but fixed, compact Hausdorff space X. The cones will have to satisfy various requirements, and we list some of these for later references:

$$\text{Every } f \in S \text{ is an u.s.c. function from } X \text{ into } [-\infty, \infty[. \tag{5.2}$$

$$\text{The constant functions belong to } S. \tag{5.3}$$

$$S \text{ separates the points of } X. \tag{5.4}$$

We shall say that S is *max-stable* if

$$f, g \in S \Rightarrow f \vee g \in S. \tag{5.5}$$

Every cone S satisfying (5.2)–(5.4) determines a pre-ordering (a reflexive and transitive relation) on $M_{\mathbb{R}}(X)$:

$$\mu \prec_S \nu \Leftrightarrow \mu(f) \leqslant \nu(f) \quad \text{all } f \in S. \tag{5.6}$$

Note that this is a proper (partial) ordering (the relation is antisymmetric) if the linear span of $C_{\mathbb{R}}(X) \cap S$ is dense in $C_{\mathbb{R}}(X)$.

Note also that

$$\mu, \nu \in M^+(X),\ \mu \prec_S \nu \Rightarrow \|\mu\| = \|\nu\|. \tag{5.7}$$

We shall use the symbol $M_{x,S}(X)$ to denote the set of all $\mu \in M_{\mathbb{R}}(X)$ such that $\varepsilon_x \prec_S \mu$, and we shall denote the set of all positive measures in $M_{x,S}(X)$ by $M_{x,S}^+(X)$. It follows from (5.7) that

$$M_{x,S}^+(X) \subset M_1^+(X) \quad \text{all } x \in X. \tag{5.8}$$

Observe that these definitions generalize previous notions, since $M_{x,S}^+(K) = M_x^+(K)$ when S is the cone $P(K)$ over a compact convex set K.

Motivated by the characterization of extreme points in Corollary I.2.4, we define a point $x \in X$ to be an *S-Choquet point* if $M_{x,S}^+(X) = \{\varepsilon_x\}$. The set of S-Choquet points is termed the *S-Choquet boundary*, and it will be denoted by the symbol $\partial_S X$. Again, $\partial_S K = \partial_e K$ if S is the cone $P(K)$ over a compact convex set K.

A closed subset F of X is said to be *S-stable* if

$$x \in F,\ \mu \in M_{x,S}^+(X) \Rightarrow \operatorname{Supp}(\mu) \subset F. \tag{5.9}$$

The S-stable sets play much the same role as the closed faces in the geometric theory. In fact, it is not hard to see that a closed subset F of a compact convex set K is $P(K)$-stable iff it is a union of faces, i.e. if face $(x) \subset F$ whenever $x \in F$. (For details cf. e.g. [5].)

We now pass to Bauer's Maximum Principle which replaces the Krein Milman Theorem in the present setup. (In fact, one may prove the classical Krein Milman Theorem by way of Bauer's principle [41].)

Theorem I.5.3. *(Bauer) If f is any function belonging to a cone S satisfying* (5.2)–(5.4), *then there exists a point $x \in \partial_S X$ such that*

$$f(x) = \sup_{y \in X} f(y). \tag{5.10}$$

Proof. Let $\beta = \sup_{y \in X} f(y) < \infty$, and define $F_0 = \{x \in X \mid f(x) = \beta\}$. By semi-continuity and compactness, F_0 is closed and non-empty. It is also easily verified that F_0 is S-stable and that the collection of S-stable (closed) subsets of F_0 is inductive with respect to inclusion. By Zorn's Lemma, F_0 contains a minimal S-stable subset F, and we claim that F must be a one-point set.

To prove this claim we assume the contrary, say $x, y \in F$ and $x \neq y$. By (5.4) there is a $g \in S$ which separates x and y, say $g(x) > g(y)$. Now we define

$$F' = \left\{ z \in F \,\middle|\, g(z) = \sup_{w \in F} g(w) = \alpha \right\},$$

and we claim that F' is an S-stable set. In fact, if $z \in F'$ and $\mu \in M^+_{z,S}(X)$, then $\operatorname{Supp}(\mu) \subset F$ by the S-stability of F, and we must also have $\operatorname{Supp}(\mu) \subset F'$ for otherwise the u.s.c. function g would be less than $g(z)-\varepsilon=\alpha-\varepsilon$ on a compact set of positive measure for some $\varepsilon>0$, which would imply $\mu(g)<g(z)$. Now F' is an S-stable subset of F excluding the point $y \in F$, contrary to the minimality of F.

Thus we have proved that F is of the form $\{x_0\}$ for some $x_0 \in F_0$, and we see that $x_0 \in \partial_S X$ since F is S-stable. □

Corollary I.5.4. *If S is any cone satisfying* (5.2)–(5.4), *then the S-Choquet boundary is non-empty.*

Every cone S satisfying (5.2)–(5.4) *generates a max-stable cone* $\tilde{S}$, consisting of all functions $f_1 \vee \cdots \vee f_n$ where $f_1, \ldots, f_n \in S$. Clearly $M^+_{x,S}(X)=M^+_{x,\tilde{S}}(X)$ for every $x \in X$. Hence we can state the following

Proposition I.5.5. *If S is a cone satisfying* (5.2)–(5.4) *then* $\partial_S X=\partial_{\tilde{S}} X$.

Henceforth we shall work only with cones of continuous functions. We shall say that a cone $S \subset C_{\mathbb{R}}(X)$ is *admissible* if it satisfies (5.3), (5.4), i.e. if it contains the constant functions and separates points.

If S is an admissible cone and $f: X \to [\alpha, \infty]$ where $\alpha \in \mathbb{R}$, then we define the *lower S-envelope* of f to be the function

$$\check{f}_S(x)=\sup\{g(x) \mid g \leqslant f, g \in S\}. \tag{5.11}$$

Similarly if $f: X \to [-\infty, \alpha]$, then we define the *upper S-envelope* of f by

$$\hat{f}_S(x)=\inf\{g(x) \mid g \geqslant f, g \in -S\}. \tag{5.12}$$

Clearly $\check{f}_S$ is l.s.c., and there is a net $\{g_\alpha\}$ from $\tilde{S}$ such that $g_\alpha \nearrow \check{f}_S$. The dual statements hold for $\hat{f}_S$. Also we can state the following simple results:

Proposition I.5.6. *The statements of Proposition* I.1.6 *can all be transferred to envelopes $\check{f}_S, \hat{f}_S$ defined by an arbitrary admissible cone.*

Proposition I.5.7. *Let S be a max-stable admissible cone and let $\mu, \nu \in M^+(X)$. Then $\mu \prec_S \nu$ iff $\nu(f) \leqslant \mu(\hat{f}_S)$ for all $f \in C_{\mathbb{R}}(C)$. In particular $\nu(f) \leqslant \hat{f}_S(x)$ for all $x \in X$, $\nu \in M^+_{x,S}(X)$ and $f \in C_{\mathbb{R}}(X)$.*

The proof of our next proposition is identical with the proof of Proposition I.3.5.

Proposition I.5.8. *Let S be an admissible cone, let $f \in C_{\mathbb{R}}(X)$, and let $\mu \in M^+(X)$ Then there exists a measure $\nu \in M^+(X)$ such that $\mu \prec_S \nu$ and $\nu(f) = \mu(\hat{f}_S)$.*

If $S \subset C_{\mathbb{R}}(X)$ is a max-stable admissible cone then $S - S$ is dense in $C_{\mathbb{R}}(X)$ by Stone's Theorem. Thus the ordering $\prec_S$ is antisymmetric. Applying the two preceding results we obtain the following generalized version of Proposition I.4.5.

Proposition I.5.9. *Let S be a max-stable admissible cone and let $\mu \in M^+(X)$. Then the following statements are equivalent:*

(i) *μ is a maximal element of $M^+(X)$ with respect to the relation $\prec_S$.*
(ii) *$\mu(\hat{f}_{\tilde{S}}) = \mu(f)$ for all $f \in C_{\mathbb{R}}(X)$.*
(iii) *$\mu(\hat{f}_S) = \mu(f)$ for all $f \in S$.*

Again we note that the equation $\mu(\hat{f}_S) = \mu(f)$ subsists for a maximal positive measure μ and an u.s.c. function f, but not for a l.s.c. function in general.

A measure $\mu \in M_1^+(X)$ with the properties (i)–(iii) above will be said to be *S-maximal.*

By Proposition I.5.5, a point $x \in X$ is an S-Choquet point iff ε_x is an $\tilde{S}$-maximal measure. Since the S- and $\tilde{S}$-envelopes coincide, we obtain the following generalized version of Proposition I.4.1:

Corollary I.5.10. *Let S be an admissible cone and let $x \in X$. Then the following statements are equivalent:*

(i) *$x \in \partial_S X$.*
(ii) *$f(x) = \hat{f}_{\tilde{S}}(x)$ for every $f \in C_{\mathbb{R}}(X)$.*
(iii) *$f(x) = \hat{f}_S(x)$ for every u.s.c. function $f: X \to [-\infty, \infty[$.*

Reversing the semi-continuity and the envelopes in the above result, we obtain the following:

Corollary I.5.11. *Let S be an admissible cone, and let $f: X \to]-\infty, \infty]$ be l.s.c. If $x \in \partial_S X$ and $f(x) > 0$, then there exists a function $g \in S$ such that*

$$g < f, \qquad g(x) > 0. \tag{5.13}$$

At this point we shall break the line of straight-forward generalizations and establish a non-trivial new result, which will be of interest also when specialized to the original geometric situation. The main advantage of the general approach rests with the following observation:

Proposition I.5.12. *If S is an admissible cone and $f: X \to [-\infty, \infty[$ is an u.s.c. function such that*

$$f(x) \leqslant \int f d\mu \tag{5.14}$$

for all $x \in X$ and all $\mu \in M^+_{x,S}(x)$, then the cone $S' = \{\lambda f + g \mid \lambda \in \mathbb{R}^+, g \in S\}$ satisfies (5.2)–(5.4) and $\partial_S X = \partial_{S'} X$.

Theorem I.5.13. *(Choquet, Edwards) If S is an admissible cone over X, then $\partial_S X$ is a Baire space in the relativized topology.*

Proof. Let $\{V_n\}_{n=1,2,\dots}$ be a sequence of subsets of $\partial_S X$ which are dense and open in the relativized topology, say $V_n = \partial_S X \cap G_n$ where G_n is an open subset of X for $n = 1, 2, \dots$. We shall prove that $\bigcap_{n=1}^{\infty} V_n$ is dense in $\partial_S X$.

To this end we consider a subset $V_0 \neq \emptyset$ of $\partial_S X$ which is open in the relativized topology, say $V_0 = \partial_S X \cap G_0$ where G_0 is an open subset of X, and we shall prove that

$$V_0 \cap \bigcap_{n=1}^{\infty} V_n \neq \emptyset. \tag{5.15}$$

Applying Corollary I.5.11 with $f = \chi_{G_0}$, we obtain an $f_0 \in S$ such that for some $x_0 \in \partial_S X$

$$\{x \mid f_0(x) \geqslant 0\} \subset G_0, \quad f_0(x_0) > 0. \tag{5.16}$$

Now we proceed by induction to construct a sequence $\{f_n\}_{n=0,1,\dots}$ from S and a sequence $\{x_n\}_{n=0,1,\dots}$ from $\partial_S X$ such that

$$\{x \mid f_n(x) \geqslant 0\} \subset G_n, \tag{5.17}$$

$$f_n(x_n) > 0, \tag{5.18}$$

$$f_n < f_{n-1} \tag{5.19}$$

for $n = 1, 2, \dots$.

We assume that the sequence has already been constructed up to f_n. By virtue of (5.18) and the density of V_{n+1} in $\partial_S X$ there exists a point $x_{n+1} \in \{x \mid f_n(x) > 0\} \cap V_{n+1}$. Now we apply Corollary I.5.11 with $\chi_{G_{n+1}} \wedge f_n$ in the place of f and x_{n+1} in the place of x to obtain a function $f_{n+1} \in S$ such that

$$\{x \mid f_{n+1}(x) \geqslant 0\} \subset G_{n+1}, \quad f_{n+1} < f_n, \quad f_{n+1}(x_{n+1}) > 0.$$

This completes the induction.

Now we define $f = \inf_n f_n$ and

$$S' = \{\lambda f + g \mid \lambda \in \mathbb{R}^+, g \in S\}.$$

By Proposition I.5.12, $\partial_{S'}X = \partial_S X$. Then by Theorem I.5.3 there is a point $y \in \partial_S X$, at which f assumes its supremum value.

By compactness of X

$$\{x \mid f(x) \geqslant 0\} = \bigcap_{n=0}^{\infty} \{x \mid f_n(x) \geqslant 0\} \neq \emptyset .$$

Hence $f(y) = \sup_{x \in X} f(x) \geqslant 0$, and it follows that $f_n(y) \geqslant 0$ for $n = 0, 1, 2, \ldots$. By (5.17) this entails $y \in G_n \cap \partial_S X = V_n$ for $n = 0, 1, 2, \ldots$, and the verification of formula (5.15) is complete. □

Corollary I.5.14. *The extreme boundary of a compact convex subset of a locally convex Hausdorff space is a Baire space in the relativized topology.*

A subset F of X is said to be a *max-boundary* for a cone $S \subset C_{\mathbb{R}}(X)$ if

$$\text{for each } f \in S \text{ there exists a point } y \in F \text{ such that } f(y) = \sup_{x \in X} f(x). \tag{5.20}$$

If $S = -S$, that is if S is a linear subspace of $C_{\mathbb{R}}(X)$, then (5.20) is equivalent to

$$\text{for each } f \in S \text{ there exists a point } y \in F \text{ such that } |f(y)| = \sup_{x \in X} |f(x)|. \tag{5.21}$$

If S admits a smallest closed max-boundary, then this set is called the *Šilov boundary* of X with respect to S.

Theorem I.5.15. *Every admissible cone S has a Šilov boundary, namely $\overline{\partial_S X}$.*

Proof. The equality (5.20) is satisfied for $F = \overline{\partial_S X}$ by virtue of Theorem I.5.3. Conversely, if F is any closed set such that (5.20) is valid, then $F \supset \partial_S X$, for if $x \in \partial_S X \setminus F$ then one could apply Corollary I.5.11 with $f = \chi_{X \setminus F}$ and obtain a contradiction. □

Remark. From Theorem I.5.15 we deduce the well known fact that $\overline{\partial_e K}$ is the Šilov boundary of a compact convex set K with respect to the space $A(K)$. This result is usually established by means of Milman's theorem:

$$F \subset K, \qquad F \text{ closed} \Rightarrow \partial_e K \cap F = \partial_e K \cap \text{cl.conv. } F. \tag{5.22}$$

(Cfr. Proof of Th. II.1.8.) Note, however, that one may as well proceed in the reverse direction, obtaining Milman's Theorem from Corollary I.5.11, which was fundamental in the proof of Theorem I.5.15.

Proposition I.5.16. *If S is an admissible cone over a metrizable compact Hausdorff space X, then there is an $f \in C_{\mathbb{R}}(X)$ such that $\partial_S X = \{x \mid \hat{f}_S(x) = f(x)\}$. In particular, one may choose f to be in the uniform closure of $\tilde{S}$.*

Proof. By the separability of $C_{\mathbb{R}}(X)$ [140] and the density of $\tilde{S} - \tilde{S}$ (Stone's Theorem), there is a dense sequence $\{g_n^1 - g_n^2\}$ in $C_{\mathbb{R}}(X)$ where $g_n^1, g_n^2 \in \tilde{S}$ for $n = 1, 2, \ldots$. We rename the functions g_n^1, g_n^2 and arrange them in a single sequence $\{f_n\}$, and we claim that the function

$$f = \sum_{n=1}^{\infty} 2^{-n} \|f_n\|^{-1} f_n$$

satisfies $\{x \mid \hat{f}_S(x) = f(x)\} \subset \partial_S X$.

To prove this claim, we consider a point $x \in X$ such that $\hat{f}_S(x) = f(x)$ and a measure $\mu \in M_{S,x}^+(X)$. Now by Proposition I.5.7

$$f(x) \leqslant \int f \, d\mu \leqslant \hat{f}_S(x) = f(x).$$

This implies $f_n(x) = \mu(f_n)$ for $n = 1, 2, \ldots$, and it follows by the density of $\{g_n^1 - g_n^2\}$ that $\mu = \varepsilon_x$, q.e.d. □

Corollary I.5.17. *If S is an admissible cone over a metrizable compact Hausdorff space X, then $\partial_S X$ is a G_δ-subset.*

Corollary I.5.18. *Let S be a max-stable admissible cone over a metrizable compact Hausdorff space X. A measure $\mu \in M_1^+(X)$ is S-maximal iff $\mu(\partial_S X) = 1$.*

Theorem I.5.19 *If S is a max-stable admissible cone over a compact Hausdorff space X, then there exists for every $x \in X$ an S-maximal measure $\mu \in M_{x,S}^+(X)$. If X is metrizable then $\mu(\partial_S X) = 1$.*

The *proof* is based on vague compactness, and it makes use of Zorn's Lemma. It is similar to the proof of Lemma I.4.7. □

Remark. By definition $\mu \in M_{x,S}^+(X)$ means that $f(x) \leqslant \mu(f)$ for all $f \in S$. If $S = -S$, that is if S is a linear space, then we shall have *equality*.

In order to treat the general, not necessarily metrizable, case, we shall have to modify the previous proof (of Prop. I.4.10), which invoked geometric notions like extreme points and faces.

Proposition I.5.20. *Let S, T be admissible cones over compact Hausdorff spaces X, Y respectively, and let φ be a continuous map of X onto Y such that $\varphi^*(T) \subset S$. Then for every point $y \in \partial_T Y$ there is a point $x \in \partial_S X$ such that $\varphi(x) = y$.*

Proof. Let $y \in \partial_T Y$ and define $X' = \varphi^{-1}(y)$ and $S' = S|X'$. Clearly S' is an admissible cone over X' and we claim that every point $x \in \partial_{S'} X'$ belongs to $\partial_S X$. To prove this claim, we consider a measure $\mu \in M^+_{x,S}(X)$. The direct image $\varphi\mu$ of μ under φ satisfies the requirement

$$\int f\, d(\varphi\mu) = \int (f \circ \varphi)\, d\mu, \quad \text{all } f \in C_{\mathbb{R}}(Y). \tag{5.23}$$

Since $f \circ \varphi \in S$ for every $f \in T$, we shall have

$$\int f\, d(\varphi\mu) = (f \circ \varphi)(x) = f(y), \quad \text{all } f \in T. \tag{5.24}$$

Since $y \in \partial_T Y$, we must have $\varphi\mu = \varepsilon_y$. Hence $\operatorname{Supp}(\mu) \subset \varphi^{-1}(y) = X'$, and so $\mu \in M^+_{x,S'}(X')$. By assumption $x \in \partial_{S'} X'$ and so $\mu = \varepsilon_x$. This completes the verification that $x \in \partial_S X$. □

Proposition I.5.21. *If S is an admissible cone and $\{f_n\}$ is an upper bounded sequence from $C_{\mathbb{R}}(X)$ such that $\limsup_n f_n(x) \leqslant \alpha$ for all $x \in \partial_S X$, then $\limsup \check{f}_{n,S}(x) \leqslant \alpha$ for all $x \in X$.*

Proof. For an arbitrary point $x_0 \in X$ there exist functions $g_n \in S$ such that

$$g_n \leqslant f_n, \qquad \check{f}_{n,S}(x_0) < g_n(x_0) + \frac{1}{n}. \tag{5.25}$$

We define $\Phi: X \to \mathbb{R}^{\aleph_0}$ by $\Phi(x) = \{g_n(x)\}$, and we write $\Phi(X) = Y$ and $\Phi(x_0) = y_0$. If $T \subset C_{\mathbb{R}}(Y)$ is the cone generated by the constant functions and the projections p_n in $\mathbb{R}^{\aleph_0}$, then $\Phi^*(T) \subset S$. By Proposition I. 5.20 there exists for every $y \in \partial_T Y$ a point $x \in \partial_S X$ such that $\Phi(x) = y$, and so

$$\limsup_n p_n(y) = \limsup_n g_n(x) \leqslant \alpha. \tag{5.26}$$

Since $M^+_{y_0,\tilde{T}}(Y) = M^+_{y_0,T}(Y)$ and $\partial_{\tilde{T}} Y = \partial_T Y$, we get by Theorem I. 5.19 that there exists a measure $\mu \in M^+_{y_0,T}(Y)$ which is concentrated on $\partial_T Y$. The sequence $\{p_n\}$ is upper bounded on $Y = \Phi(X)$ since $\{g_n\}$

is upper bounded on X. Since the point y in (5.26) was an arbitrary point of $\partial_T Y$, we may apply Fatou's Lemma as follows:

$$\begin{aligned}\limsup_n \check{f}_{n,S}(x_0) &= \limsup_n g_n(x_0) = \limsup_n p_n(y_0) \\ &\leq \limsup_n \mu(p_n) \leq \mu\left(\limsup_n p_n\right) \leq \alpha .\end{aligned}$$

This completes the proof. □

Now the proofs of our next two results are similar to those of Corollary I. 4.12 and Theorem I. 4.14.

Proposition I.5.22. *If S is a max-stable admissible cone and $\mu \in M_1^+(X)$ is an S-maximal measure then $\mu(C)=0$ for every compact G_δ-set C disjoint from $\partial_S X$.*

Theorem I.5.23. *If S is an admissible cone over a compact Hausdorff space X, then there exists for every point $x \in X$ a positive countably additive set-function $\bar{\mu}$ on the σ-field $\partial_S X \cap \mathscr{B}_0$ such that*

$$f(x) \leq \int_{\partial_S X} f \, d\bar{\mu} \quad \text{all } f \in S \tag{5.27}$$

and

$$\bar{\mu}(\partial_S X) = 1 . \tag{5.28}$$

Final remark. The definition of *Choquet boundary* can be modified slightly such as to apply to cones $S \subset C_{\mathbb{R}}(X)$ which do not separate points. If ρ is the equivalence relation

$$x \sim y \Leftrightarrow f(x) = f(y) \quad \text{for all } f \in S ,$$

and if φ is the cannonical mapping of X onto $Y = X/\rho$, then the *saturation of a point* x is the closed set $\varphi^{-1}\varphi(x)$, and a point x is said to be a *Choquet point* for S if

$$\varepsilon_x \prec_S \mu, \ \mu \in M_1^+(X) \Rightarrow \mu(\varphi^{-1}\varphi(x)) = 1 .$$

However, every $f \in S$ can be factorized as $f = \tilde{f} \circ \varphi$, and it is easily seen that x is a Choquet point of X with respect to S iff $\varphi(x)$ is a Choquet point of Y with respect to the point-separating cone of all $\tilde{f}$ where $f \in S$. (Y is endowed with quotient topology and it is observed that Y becomes a compact Hausdorff space.)

Notes and remarks. Proposition I.5.1 was proved by G. Mokobodzki in 1962 [297], and the maximum principle of Theorem I.5.3 was proved by H. Bauer in 1960 [39]. Theorem I.5.13 is fairly recent, but it has an interesting pre-history. In 1951 I. Kaplansky proved that the structure spaces of certain operator algebras (CCR-algebras) are Baire spaces [236]. This result was extended to arbitrary C^*-algebras by 1960 in Dixmier [137]. He used it to show that the kernel of a factor representation of a seprable C^*-algebra must also be the kernel of an irreducible representation. The fact that extreme boundaries of compact convex sets are Baire spaces, was proved by G. Choquet. He did not publish his result, which first appeared in 1964 in an appendix to Dixmier's book on C^*-algebras [138]. In Dixmier's book it is also shown how this theorem of Choquet yields Kaplansky's result for completely general C^*-algebras. Cf. also the historical note to Ch. II, § 6. The generalization from extreme boundaries to abstract Choquet boundaries was given by D.A. Edwards in 1966 [146].

We remark that the definitions of Choquet boundary and max-boundary generalize in a natural way to a family S (not necessarily a cone) of u.s.c. functions on X.

If S is a family of complex valued functions we define a subset F of X to be a max-boundary (Šilov boundary) for S if F is a max-boundary (Šilov boundary) for the real valued family $|S|=\{|f| \mid f\in S\}$. The existence of a Šilov boundary was proved first for a point-separating *algebra* of continuous complex valued functions by G. Šilov in 1940; but the publication of his result was delayed till 1946 because of the war [183]. In 1954 Šilov's Theorem was generalized by R. Arens and I.M. Singer to the case of a multiplicative semigroup of continuous functions [24]. The corresponding result for admissible (in the sense of § 5) linear subspaces of $C_{\mathbb{R}}(X)$ and $C_{\mathbb{C}}(X)$ was announced by H. Bauer in 1958 [37]. The same result was proved independently by E. Bishop and [K. de Leeuw in 1959 [63], and by H. S. Bear in 1960 [49]. In the proofs of Šilov, Arens-Singer and Bear, the existence of a *minimal* closed max-boundary F is established by Zorn's Lemma; then it is shown by a separate argument that F must actually be the *smallest* closed max-boundary. The connection with the Choquet boundary was first established in the works of Bauer and of Bishop-de Leeuw. In Bishop-de Leeuw's paper there is also a stronger result valid for the case in which A is a norm-closed and point-separating subalgebra A of $C_{\mathbb{C}}(X)$, where each point of X is a G_δ. "For every $x\in\partial_A X$ there exists an $f\in A$ which "peaks" at x, in that $\{x\}=\{y \mid |f(y)|=\|f\|_\infty\}$; a fortiori $\partial_A X$ is the smallest (not necessarily closed) max-boundary for A." (For this reason $\partial_A X$ is often termed "the minimal boundary" in this case). We remark that by definition $\partial_A X$ is equal to $\partial_{\mathrm{Re}\,A} X$, where $\mathrm{Re}\,A$ is the linear space of real parts of functions in A. A slightly more general theorem of the same kind was proved by J. Siciak in 1962 [368]. He proves that if A is a point-separating family of non-negative functions on a compact metric space X and A is closed under multiplication, then E has a (not necessarily closed) smallest max-boundary.

The most general known theorem on the existence of a Šilov boundary is given in H. Bauer's paper from 1961 [40]. Here he proves that the Choquet boundary of a family $\mathscr{E}$ of u.s.c. functions on a compact Hausdorff space X such that $\mathscr{E}+\mathscr{E}\subset\mathscr{E}$, is a max-boundary for $\mathscr{E}$ contained in any *closed* max-boundary for $\mathscr{E}$; hence $\overline{\partial_{\mathscr{E}} X}$ is the Šilov boundary for $\mathscr{E}$. His proof is essentially the same as the one given in the preceding § 5, but the absence of scalars forces him to replace the application of the Hahn-Banach Theorem by an application of Aumann's Theorem on the extension of additive real functions on Abelian groups dominated by subadditive functionals [33]. (Note that one may pass from the multiplicative setting of Arens-Singer to the additive setting by taking logarithms.) An excellent survey on Šilov boundary theory is given in [44].

The generalization of the boundary measure representation theory from the extreme boundary of a compact convex set to abstract Choquet boundaries, is the work of many hands. Besides the fundamental works of Bishop-de Leeuw and Bauer, mentioned above, some relevant recent references are: D.A. Edwards [148], E.B. Davies [124], and N. Boboc and A. Cornea [72].

§ 6. Unilateral Representation Theorems with Application to Simplicial Boundary Measures

We shall use the term *ordered convex compact* to denote a compact convex subset K of a locally convex Hausdorff space E over $\mathbb{R}$ provided with a (partial) ordering defined by a closed (and proper) cone E^+. A real valued function l on K is said to be *isotone* if $l(x)\leqslant l(y)$ whenever $x\leqslant y$. The convex cone of isotone functions in $A(K)$ (respectively $A(K;E)$) is denoted by $L(K)$ respectively $L(K;E)$).

Proposition I.6.1. *If x, y are points in an ordered convex compact K, then*

$$x\leqslant y \Leftrightarrow l(x)\leqslant l(y) \quad \text{all } l\in L(K). \tag{6.1}$$

In particular, $L(K)$ separates the points of K.

Proof. To prove the non-trivial part of the equivalence (6.1), we assume $x\nleqslant y$. Now $y-x\notin E^+$, and by Hahn-Banach separation there is a $q\in E^*$ such that

$$q(y-x)<\inf\{q(z)|z\in E^+\}.$$

In particular q is bounded below over the cone E^+, and so $q(E^+)\subset \mathbb{R}^+$. It follows that $q|K\in L(K)$ and that $q(y)<q(x)$. This completes the proof of (6.1).

Finally assume $x\neq y$, say $x\nleq y$. By the first part of the proof there is an $l\in L(K)$ such that $l(y)<l(x)$. Hence $L(K)$ separates points. ▯

Remark. The cone $L(K)$ may be replaced by the cone $L(K;E)$ in Proposition I.6.1.

The set of points of an ordered convex compact K which are *maximal* in the given ordering, will be denoted by $Z(K)$, or briefly by Z.

Proposition I.6.2. *If K is an ordered convex compact, then there exists for every point x in K a maximal point z in K such that $x\leqslant z$. In particular $Z\neq\emptyset$. Moreover, Z is a union of faces.*

Proof. 1. The set $\{y\in K|x\leqslant y\}$ is inductive by the compactness of K and the fact that E^+ is closed. By Zorn's Lemma it has a maximal element.

2. If $x\in Z$ and $x=\lambda y+(1-\lambda)z$ where $y,z\in K$ and $0<\lambda<1$, then $y,z\in Z$; for if $y<y'\in K$, say, then $x<\lambda y'+(1-\lambda)z\in K$ contrary to the maximality of x. Hence $\operatorname{face}(x)\subset Z$ for every $x\in Z$, q.e.d. ▯

Corollary I.6.3. *If Z is convex, then it is a face of K.*

It follows from Proposition I.6.2 that a point z in Z is an extreme point of K iff

$$z=\tfrac{1}{2}x+\tfrac{1}{2}y; \quad x,y\in Z \Rightarrow x=y. \tag{6.2}$$

If Z is convex, then (6.2) simply means that $z\in\partial_e Z$. For the sake of brevity we shall write $\partial_e Z$ in the place of $Z\cap\partial_e K$ also in the general case.

It is not entirely obvious that the set $\partial_e Z$ of an ordered convex compact K is non-empty, since Z may fail to be compact. Thus, instead of using the Krein-Milman Theorem, we shall have to give a separate existence theorem based on Zorn's Lemma.

A subset F of K is said to be *hereditary upwards* if

$$x\in F, \quad y\in K, \quad x\leqslant y \Rightarrow y\in F \tag{6.3}$$

Proposition I.6.4. *If a closed face F_0 of an ordered convex compact K is hereditary upwards, then $F_0\cap\partial_e Z\neq\emptyset$. In particular $\partial_e Z\neq\emptyset$.*

Proof. Let $\mathscr{F}$ be the collection of all closed faces contained in F_0 which are hereditary upwards. Clearly $\mathscr{F}$ is inductive in the ordering $\supset$, and by Zorn's Lemma it has a minimal element F. Now F must be a one-point set, for if $x, y \in F$ and $x \neq y$, then the two points x, y could be separated by a function $l \in L(K)$, and then the set

$$F' = \left\{ z \in F \,\middle|\, l(z) = \sup_{w \in F} l(w) \right\}.$$

would belong to $\mathscr{F}$ and be properly contained in F, contrary to the minimality.

Thus we shall have $F = \{x\}$ for some $x \in F_0$. The point x must be an extreme point of K since F is a face, and x must be a maximal point since F is hereditary upwards. This completes the proof. □

The $L(K)$-envelopes of functions on an ordered convex compact K will be called *monotone* convex and concave envelopes, and they will be denoted as follows:

$$\check{\check{f}} = \check{f}_{L(K)}, \qquad \hat{\hat{f}} = \hat{f}_{L(K)}. \tag{6.4}$$

Clearly, $\check{\check{f}}$ and $\hat{\hat{f}}$ are related to the usual convex and concave envelopes by the formula

$$\check{\check{f}} \leqslant \check{f} \leqslant f \leqslant \hat{f} \leqslant \hat{\hat{f}}. \tag{6.5}$$

The ordering defined on $M^+(K)$ by the max-stable cone $\tilde{L}(K)$ generated by $L(K)$, will be denoted by the symbol $\prec\!\!\prec$. Thus

$$\mu \prec\!\!\prec \nu \;\Leftrightarrow\; \mu(f) \leqslant \nu(f) \quad \text{all} \quad f \in \tilde{L}(K). \tag{6.6}$$

Observe that the ordering $\prec\!\!\prec$ is anti-symmetric since $L(K)$, and hence $\tilde{L}(K)$, separates points by Proposition I.6.1. Clearly this ordering is related to the ordering of Choquet by the formula

$$\mu \prec \nu \Rightarrow \mu \prec\!\!\prec \nu. \tag{6.7}$$

Observe also that if $\mu, \nu \in M_1^+(K)$ and $\mu \prec\!\!\prec \nu$, then for every function $l \in L(K) \subset A(K)$:

$$l(x_\mu) = \mu(l) \leqslant \nu(l) = l(x_\nu).$$

Hence by Proposition I.6.1:

$$\mu, \nu \in M_1^+(K), \qquad \mu \prec\!\!\prec \nu \;\Rightarrow\; x_\mu \leqslant x_\nu. \tag{6.8}$$

Proposition I.6.5. *A point x of an ordered convex compact K belongs to Z iff $a(z) = \hat{\hat{a}}(z)$ for all $a \in A(K)$.*

Proof. 1. Assume first that $z \in Z$, and consider an arbitrary $a \in A(K)$. By Proposition I.5.8 there exists a $\mu \in M_1^+(K)$ such that $\varepsilon_z \lll \mu$ and $\mu(a) = \hat{a}(z)$. By virtue of (6.8) and by the maximality of z, we shall have $x_\mu = z$. Since $a \in A(K)$, this yields $a(z) = \mu(a) = \hat{a}(z)$.

2. Assume next that $z \notin Z$, say $z < y \in K$. By Proposition I.6.1, there is an $l \in L(K)$ such that $l(z) < l(y)$. Now

$$l(z) < l(y) \leqslant \hat{\hat{l}}(y) \leqslant \hat{\hat{l}}(z),$$

and the proof is complete. □

Theorem I.6.6. *If x is a point of an ordered convex compact, then the following statements are equivalent*

(i) $x \in \partial_e Z$,

(ii) $f(x) = \hat{\hat{f}}(x)$ *for all* $f \in C_{\mathbb{R}}(K)$,

(iii) x *belongs to the* $L(K)$*-Choquet boundary of* K.

Proof. (i) ⇒ (ii). We assume (i) and consider an aribitrary $f \in C_{\mathbb{R}}(K)$. By Hervé's Theorem (Prop. I.4.1) there is an $a \in A(K)$ such that

$$f \leqslant a, \qquad a(x) < f(x) + \frac{\varepsilon}{2}, \tag{6.9}$$

where ε is an arbitrary positive number.

By Proposition I.6.5 we shall have $\hat{a}(x) = a(x)$, and the definition of monotone envelopes yields an $l \in -L(K)$ such that

$$a \leqslant l, \qquad l(x) < a(x) + \frac{\varepsilon}{2}. \tag{6.10}$$

Combining (6.9) and (6.10) we obtain $f \leqslant l$ and $l(x) < f(x) + \varepsilon$. It follows that $\check{f}(x) = f(x)$, q.e.d.

(ii) ⇒ (iii). This is a direct consequence of Corollary I.5.10.

(iii) ⇒ (i). We assume that $x \in \partial_{L(K)} K$. To prove $h \in Z$, we assume $x \leqslant y \in K$. Every function in $\tilde{L}(K)$ is isotone, and so $l(x) \leqslant l(y)$ for all $l \in \tilde{L}(K)$. By definition of the ordering $\lll$, this gives $\varepsilon_x \lll \varepsilon_y$. Now we shall have $x = y$ since x is a Choquet point, and the maximality is proved.

To prove $x \in \partial_e K$, we consider an arbitrary $\mu \in M_x^+(K)$. Now $\varepsilon_x \prec \mu$, and by (6.7) $\varepsilon_x \lll \mu$. Again it follows that $\varepsilon_x = \mu$ since x is a Choquet point. Hence x is an extreme point of K (Cor. I.2.4), and the proof is complete. □

Applying Theorem I.5.13, Corollary I.5.17 and Theorem I.5.23, we obtain the following

Corollary I.6.7. *If K is an ordered convex compact, then then $\partial_e Z$ is a (non-empty) Baire space in the relativized topology, and for every $x \in K$ there exists a positive and countably additive set function $\bar{\mu}$ on the σ-field $\partial_e Z \cap \mathscr{B}_0$ such that*

$$l(x) \leqslant \int_{\partial_e Z} l \, d\bar{\mu} \quad \text{all } l \in L(K) \tag{6.11}$$

and

$$\bar{\mu}(\partial_e Z) = 1 \, . \tag{6.12}$$

If K is metrizable, then $\partial_e Z$ is a G_δ-subset, and $\bar{\mu}$ becomes a Baire-measure.

It is particularly interesting to study the case in which the set Z of maximal points of K is convex. Then Z is a face of K (Cor. I.6.3), and it will be called the *positive face* of K.

It follows from the above results that if K is an ordered convex compact possessing a positive face Z, then the Choquet-Bishop-de Leeuw Theorem will be valid for the (possibly non-compact) convex set Z. Specifically we shall have the following:

Corollary I.6.8. *If K is an ordered convex compact with a positive face Z, then there exists for every point $x \in Z$ a positive and countably additive set function $\bar{\mu}$ on the σ-field $\partial_e Z \cap \mathscr{B}_0$ such that*

$$a(x) = \int_{\partial_e Z} a \, d\bar{\mu} \quad \text{all } a \in A(K) \, , \tag{6.13}$$

and

$$\bar{\mu}(\partial_e Z) = 1 \, . \tag{6.14}$$

Proof. The (Radon-)measure μ associated with the set function $\bar{\mu}$ of Corollary I.6.7, satisfies $\varepsilon_x \prec\!\!\prec \mu$. By virtue of (6.8) this implies $x \leqslant x_\mu$, and (6.13) follows. □

We shall close this paragraph with an application yielding representing boundary measures that are "irredundant" or "simplicial". Specifically, we shall say that a positive and normalized measure μ on a compact convex set K is *simplicial* if μ is an extreme point of $M^+_{x_\mu}(K)$.

Proposition I.6.9. *Let K be a compact convex set and let $\mu \in M^+_1(K)$. Then μ is simplicial iff $A(K)$ is dense in $L^1(\mu)$.*

Proof. 1. Let $x=x_\mu$, and assume first that $A(K)$ is non-dense in $L^1(\mu)$. Then there is a non-zero element h of $L^\infty(\mu)$ such that $\|h\|_\infty \leqslant 1$ and

$$\int a h \, d\mu = 0, \quad \text{all } a \in A(K). \tag{6.15}$$

Define the measure ν by $d\nu = h\,d\mu$. By definition $-\mu \leqslant \nu \leqslant \mu$. Hence the two measures $\mu_1 = \mu + \nu$, $\mu_2 = \mu - \nu$ are positive and non-equal. Application of (6.15) gives

$$\mu_i(a) = \mu(a) = a(x), \quad \text{all } a \in A(K); \quad i = 1,2. \tag{6.16}$$

Hence $\mu_1, \mu_2 \in M_x^+(K)$. Now μ is non-extreme, since $\mu = \frac{1}{2}\mu_1 + \frac{1}{2}\mu_2$.

2. Assume next that μ is a non-extreme point of $M_x^+(K)$, say $\mu = \frac{1}{2}\mu_1 + \frac{1}{2}\mu_2$ where $\mu_1, \mu_2 \in M_x^+(K)$ and $\mu_1 \neq \mu_2$. Now $0 \leqslant \mu_1 \leqslant 2\mu$. Hence $d\mu_1 = h\,d\mu$ where $0 \leqslant h \leqslant 2$ a.e. (μ). Also $1-h$ is a non-zero element of $L^\infty(\mu)$ since $\mu_1 \neq \mu_2$.

Now we shall have

$$\int a(1-h)\,d\mu = \int a\,d\mu - \int a\,d\mu_1 = 0,$$

for all $a \in A(K)$. Hence $A(K)$ is non-dense in $L^1(\mu)$, and the proof is complete. □

Proposition I.6.10. *Let K be a compact convex set and let $\mu \in M_1^+(K)$ be a simple measure, say $\mu = \sum_{j=1}^{n} \lambda_j \varepsilon_{x_j}$ where $0 < \lambda_j$ for $j = 1, \ldots, n$ and $\sum_{j=1}^{n} \lambda_j = 1$. Then μ is simplicial iff $\{x_1, \ldots, x_n\}$ is an affinely independent set of points.*

Proof. 1. Let $x = x_\mu = \sum_{j=1}^{n} \lambda_j x_j$ and assume first that μ is a non-extreme point of $M_x^+(K)$, say $\mu = \frac{1}{2}\mu_1 + \frac{1}{2}\mu_2$ where $\mu_1, \mu_2 \in M_x^+(K)$ and $\mu_1 \neq \mu_2$.

Necessarily

$$\mu_i = \sum_{j=1}^{n} \alpha_{ij} \varepsilon_{x_j},$$

where $0 \leqslant \alpha_{ij} \leqslant 2\lambda_j$ for $j = 1, \ldots, n$ and $i = 1,2$.

Also we shall have $x = \sum_{j=1}^{n} \alpha_{ij} x_j$ for $i = 1,2$, and so

$$\sum_{j=1}^{n} (\alpha_{1j} - \alpha_{2j}) x_j = 0. \tag{6.17}$$

The coefficients $\alpha_{1j}-\alpha_{2j}$ have zero sum, and they do not all vanish since $\mu_1 \neq \mu_2$. Hence (6.17) is an affine dependence.

2. Assume next that $\{x_1, \dots, x_n\}$ is an affinely dependent set of points, say

$$\sum_{j=1}^{n} \beta_j x_j = 0, \qquad \sum_{j=1}^{n} \beta_j = 0, \tag{6.18}$$

where the coefficients $\beta_1, \dots, \beta_n$ are not all zero.

Define $h \in L^\infty(\mu)$ as follows:

$$h(y) = \begin{cases} \lambda_j^{-1}\beta_j & \text{if } y = x_j, \quad \text{for } j = 1, \dots, n, \\ 0 & \text{if } y \notin \{x_1, \dots, x_n\}. \end{cases}$$

Clearly h is a non-zero element of $L^\infty(\mu)$, and

$$\int a h \, d\mu = \sum_{j=1}^{n} \beta_j a(x_j) = a\left(\sum_{j=1}^{n} \beta_j x_j\right) - a(0) = 0$$

for all $a \in A(K; E)$, and hence for all $a \in A(K)$ by Corollary I.1.5.

It follows that $A(K)$ is non-dense in $L^1(\mu)$. Hence μ is non-simplicial by Proposition I.6.9. □

Proposition I.6.11. *Let K be a compact convex set in $\mathbb{R}^n$ and let $\mu \in M_1^+(K)$. Then μ is simplicial iff μ is supported by an affinely independent set of (at most $n+1$) points of $\mathbb{R}^n$.*

Proof. The "if" part is trivial by Proposition I.6.10. To prove the "only if" part, we assume that μ is simplicial. Let F be the affine span of $\operatorname{Supp}(\mu)$ and set $k = \dim(F)$ and $K' = K \cap F$. Let $J = \{x_0, \dots, x_k\}$ be an affinely independent set of points from $\operatorname{Supp}(\mu)$ whose affine span is F. We assume that there is a point $y \in \operatorname{Supp}(\mu) \backslash J$, and we shall obtain a contradiction.

Let U be a bounded open neighbourhood of 0 in $\mathbb{R}^n$ such that the sets

$$x_0 + U, \dots, x_k + U; \qquad y + U$$

are mutually disjoint. Since μ is a simplicial measure, there exists by Proposition I.6.9 a sequence $\{a_m\}$ of (continuous) affine functions on F such that

$$\int |\chi_{(y+U)\cap K'} - a_m| \, d\mu \to 0. \tag{6.19}$$

In particular, we must have

$$\int_W |a_m| \, d\mu \to 0, \tag{6.20}$$

where $W = \bigcup_{j=0}^{k} (x_j + U) \cap K'$.

The functional $p: a \rightsquigarrow \int_W |a| d\mu$ is a semi-norm on $A(F)$. In fact it is a norm, for if $a \in A(F)$ and $p(a)=0$, then

$$a(x_0) = \cdots = a(x_k) = 0,$$

since a is continuous and $x_i \in \mathrm{Supp}(\mu)$ for $i=0, \ldots, k$. Now it follows that $a \equiv 0$ on F, since a is affine and F is spanned by $x_0, \ldots, x_k$.

The space $A(F)$ is of finite dimension $(k+1)$, and so the norm p is topologically equivalent to the supremum norm over $K' = K \cap F$. Hence it follows from (6.20) that $a_m \to 0$ uniformly on $K \cap F$. By virtue of (6.19) and the fact that $\mathrm{Supp}(\mu) \subset K \cap F$, this implies $\mu[(y+U) \cap K] = 0$. This gives the desired contradiction since $y \in \mathrm{Supp}(\mu)$ and U is open. □

Henceforth we shall denote the set of all positive and normalized boundary measures representing a point x in a compact convex set K, by the symbol $Z_x(K)$ or simply by Z_x. Clearly $M_x^+(K)$ is an ordered convex compact in the ordering of Choquet and in the vague topology, and Z_x is the set of maximal elements of $M_x^+(K)$. Moreover

$$Z_x = \{\mu \in M_x^+(K) \mid \mu(\hat{f} - f) = 0 \text{ all } f \in C_{\mathbb{R}}(K)\} \tag{6.21}$$

by virtue of Proposition I.4.5, and it follows that Z_x is a *face* of $M_x^+(K)$. In the language of ordered convex compacts, Z_x is the *positive face* of $M_x^+(K)$. Note that Z_x will not be closed in general.

The elements of

$$\partial_e Z_x = \partial_e M_x^+(K) \cap Z_x$$

are the *simplicial boundary measures which represents* x. In the remaining part of this paragraph we shall show that there are "sufficiently many" such measures for every point x in a compact convex set K.

Proposition I.6.12. *If x is a point of a compact convex set K and if $f \in P(K)$, then is a $\mu \in \partial_e Z_x$ such that $\mu(f) = \hat{f}(x)$. In particular $\partial_e Z_x \neq \emptyset$.*

Proof. The set $F_0 = \{\mu \in M_x^+(K) \mid \mu(f) = \hat{f}(x)\}$ is nonempty by Corollary I.3.6. It is seen to be a closed face which is hereditary upwards. Hence it follows from Proposition I.6.4 that $F_0 \cap \partial_e Z_x \neq \emptyset$, q.e.d. □

Corollary I.6.13. *(Carathéodory) Every point x of a compact convex subset K of $\mathbb{R}^n$ is a convex combination of (at most $n+1$) affinely independent extreme points.*

Proof. Combination of the two preceding propositions and the fact that a simple measure $\mu = \sum_{i=1}^{n} \mu_i \varepsilon_{x_i}$ where all $\lambda_i \neq 0$ is a boundary measure iff $x_i \in \partial_e K$ for $i = 1, \dots, n$ (cf. Ch. I, § 4). □

If K is a compact convex set, then we shall use the symbol $D(K)$ to denote the linear span of $C_{\mathbb{R}}(K)$ and $\{\hat{f} \mid f \in C_{\mathbb{R}}(K)\}$ in the vector space of all bounded Borel functions on K. Note that $\mu(f)$ is well defined for every $f \in D(K)$ and every $\mu \in M_{\mathbb{R}}(K)$.

Theorem I.6.14. *Let x be an arbitrary point of a compact convex set K. Then Z_x is the closed convex hull of $\partial_e Z_x$ in the weak topology σ defined on $M_{\mathbb{R}}(K)$ by $D(K)$.*

Proof. It follows from (6.21) that Z_x is closed in the topology σ. To prove that $Z_x \subset \sigma\text{-cl.conv}(\partial_e Z_x)$, we assume the contrary, say

$$\mu \in Z_x \backslash \sigma\text{-cl.conv}(\partial_e Z_x)\,. \tag{6.22}$$

By Hahn-Banach separation there is a $k \in D(K)$ such that

$$\sup_{\nu \in \partial_e Z_x} \nu(k) = \alpha < \mu(k)\,. \tag{6.23}$$

By definition $k = f_0 + \sum_{i=1}^{n} \beta_i \hat{f}_i$, where $f_i \in C_{\mathbb{R}}(K)$ for $i = 0, \dots, n$ and where $\beta_1, \dots, \beta_n \in \mathbb{R}$. All measures involved in (6.23) are boundary measures, and therefore one may replace k by the function $k' = f_0 + \sum_{i=1}^{n} \beta_i f_i \in C_{\mathbb{R}}(K)$ (cf. Prop. I.4.5). Thereupon one may replace k' by $\check{k}'$, and now it follows from the definition of lower envelopes together with a simple limit argument (cf. formula (2.3)), that there exists a function $g \in P(K)$ such that $g \leqslant k'$ and such that

$$\sup_{\nu \in \partial_e Z_x} \nu(g) \leqslant \alpha < \mu(g)\,. \tag{6.24}$$

By Proposition I.6.12 this gives a contradiction, since $\nu(g) = \hat{g}(x) \geqslant \mu(g)$ for some measure $\nu \in \partial_e Z_x$. □

Remark. The conclusion of Theorem I.6.14 does not subsist if σ is replaced by the vague topology, or by the norm topology, of $M_{\mathbb{R}}(K)$. In fact let $K = \text{conv}(D \cup L)$ where D is a plane disk in $\mathbb{R}^3$ and L is a line segment orthogonal to D which meets D in a point $y \in \partial_e D \cap (L \backslash \partial_e L)$

("Bourbaki Example" [85; Ch. II, § 7, ex. 11]). Now the set Z_x determined by the centre x of D is non-closed in the vague topology; whereas $\partial_e Z_x$ consists of measures supported by at most four points, so that the norm-closed convex hull of $\partial_e Z_x$ will consist of discrete measures only. Hence

$$\text{norm-cl.conv}(\partial_e Z_x) \subset Z_x \subset w^*\text{-cl.conv}(\partial_e Z_x),$$

and both inclusions are *strict*.

Theorem I.6.15. *For every point x of a compact convex set K the set $\partial_e Z_x$ is a Baire set in the relativized vague topology. If K is metrizable, then $\partial_e Z_x$ is a vague G_δ-subset of $M_x^+(K)$.*

Proof. Apply Corollary I.6.7 and observe that metrizability of K implies separability of $C_{\mathbb{R}}(K)$ and vague metrizability of $M_1^+(K)$. □

Finally we shall show that every boundary measure representing a point x, can be "decomposed" into simplicial boundary measures representing x.

Theorem I.6.16. *Let x be a point of a compact convex set K and let $\mathscr{B}_0(M_x^+(K))$ be the σ-field of vague Baire subsets of $M_x^+(K)$. Then there exists for every $\mu \in Z_x$ a positive and countably additive set-function θ on the σ-field $\partial_e Z_x \cap \mathscr{B}_0(M_x^+(K))$ such that*

$$\mu(f) = \int_{\partial_e Z_x} v(f)\, d\theta(v), \quad \text{all } f \in C_{\mathbb{R}}(K), \tag{6.25}$$

and

$$\theta(\partial_e Z_x) = 1. \tag{6.26}$$

Proof. Apply Corollary I.6.8 and the fact that $\mu \rightsquigarrow \mu(f)$ is a vaguely continuous affine function on $M_x^+(K)$ for every $f \in C_{\mathbb{R}}(K)$. □

Notes and remarks. The study of ordered convex compacts was initiated in 1963, by G. Lumer, who proved the basic existence theorem (Prop. I.6.4) for maximal extreme points [283]. The characterization of $\partial_e Z_x$ and the consequences hereof (Th. I.6.6, Cor. I.6.7, Cor. I.6.8) seem to have appeared rather recently in connection with the generalization of Carathéodory's Theorem by means of simplicial boundary measures (see below).

An excellent bibliography on the classical Carathéodory Theorem and related matters can be found in the survey article [121] by

Dantzer, B. Grünbaum, and V. Klee. A possible rephrasing of Carathéodory's Theorem is the following: "To every point x in a compact convex subset K of $\mathbb{R}^n$ there exists a simplex Δ such that $x \in \Delta \subset K$ and $\partial_e \Delta \subset \partial_e K$." Here the conclusion no longer invokes the dimension of the space, and so it is natural to ask if it will subsist in general when Δ is allowed to be a general (Choquet) simplex? (For the definition of Choquet simplexes cf. Chap. II.) A counterexample to this conjecture was given in 1963 by E. M. Alfsen [5].

Proposition I.6.9 was proved in 1964 by R. G. Douglas [139], while the non-vanishing of $\partial_e Z_x$ (prop. I.6.12) was established in 1968 by F. G. Vincent-Smith (presented in a lecture at the Danish Math. Soc. and published later in [394]). Vincent-Smith's proof is based on a general method to obtain Hahn-Banach extensions satisfying certain additional requirements. Essentially the same technique has been developed by P.R. Andenæs [20]. The proof presented here, and also the application to Carathéodory's Theorem (Cor. I.6.13), was given independently by G. Mokobodzki and M. Rogalski at approximately the same time (unpublished). Another proof based on Bauer's maximum principle was given at about the same by E.M. Alfsen and C. Skau [17]. Here they also established Theorem I.6.14. Finally, Theorem I.6.15 and Theorem I.6.16, are due to Vincent-Smith [394] and to Alfsen-Skau [17].

Chapter II

Structure of Compact Convex Sets

§ 1. Order-unit and Base-norm Spaces

A (partially) ordered vector space A (over $\mathbb{R}$) is said to be *Archimedean* if the negative elements $a \in A^-$ are the only ones for which $\{\alpha a \mid \alpha \in R^+\}$ has an upper bound. A vector subspace J of ordered vector space is said to be an *order ideal* if

$$a, b \in J, \quad c \in A, \quad a \leqq c \leqq b \Rightarrow c \in J. \tag{1.1}$$

Proposition II.1.1. *Let J be a vector subspace of an ordered vector space A, and let $\varphi: A \to A/J$ be the canonical homomorphism. The cone $\varphi(A^+)$ is proper iff J is an order ideal.*

Proof. 1. Let J be an order ideal, and consider an equivalence class $H \in \varphi(A^+) \cap -\varphi(A^+)$. We shall prove that $H = J$.

By assumption there are elements $a_1, a_2 \in A^+$ such that $a_1, -a_2 \in H$, and so $0 \leqq a_1 \leqq a_1 + a_2 \in J$. It follows that $a_1 \in J$, and so $H = J$.

2. Let $\varphi(A^+)$ be a proper cone, and assume that $a \leqq c \leqq b$ where $a, b \in J$. Now $\varphi(c) \in \varphi(a) + \varphi(A^+) = \varphi(A^+)$ and $\varphi(c) \in \varphi(b) - \varphi(A^+) = -\varphi(A^+)$. Hence $\varphi(c) \in \varphi(A^+) \cap -\varphi(A^+) = (0)$, and so $c \in J$, which proves that J is an order ideal. □

By Proposition II. 1.1, the quotient space A/J of an ordered vector space A modulo an order ideal J is an ordered vector space with positive cone $\varphi(A^+)$. This ordering of A/J is called the *quotient ordering.*

Remark. The quotient A/J of an Archimedean ordered vector space A modulo an order ideal J need not be Archimedean. As an example one may consider $A = \mathbb{R}^3$ with positive cone A^+ generated by the set $\{(x, y, 1) \mid (x, y) \in K\}$, where K is the closed convex set shown in the diagram. It is easily verified that A^+ is a convex cone defining an Archimedean ordering. Also it is easily verified that $J = \{(0, 0, z) \mid z \in \mathbb{R}\}$ is an order ideal, and that the quotient A/J is (isomorphic to) the xy-plane with positive cone generated by K. More specifically, this cone consists of the open right halfplane plus the upper half of the y-axis. It defines the *lexicographic* ordering of $\mathbb{R}^3$, which is seen to be non-Archimedean.

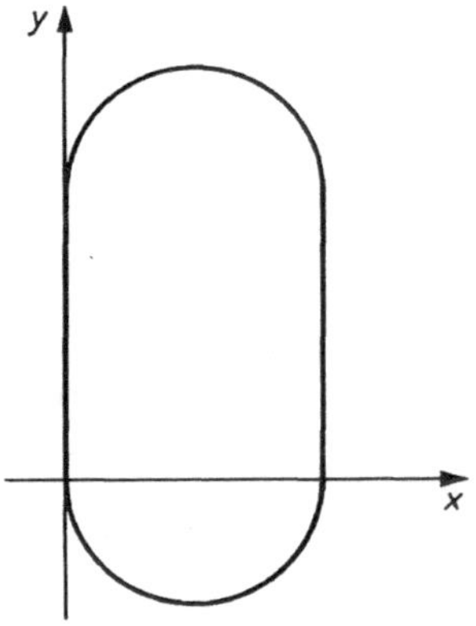

Fig. 1

If a is an element of an ordered vector space A, then the smallest order ideal containing a will be called the *order ideal generated by a*, and it will be denoted by the symbol $J(a)$.

Clearly, for every $a \in A^+$

$$b \in J(a) \Leftrightarrow -na \leqq b \leqslant na \quad \text{for some } n = 1, 2, \ldots .$$

A positive element e of an ordered vector space A is said to be an *order unit* if $J(e) = A$. It is easily verified that an ordered vector space A with an order unit e is *Archimedean iff*

$$na \leqq e \quad \text{all } n = 1, 2, \ldots \Rightarrow a \leqq 0 . \tag{1.2}$$

Observe that if an ordered vector space A admits an order unit, then A is *positively generated* (directed), in that $A = A^+ - A^+$.

Proposition II.1.2. *An Archimedean ordered vector space A with an order unit e admits a norm*

$$\|a\| = \inf\{\lambda > 0 \mid -\lambda e \leqq a \leqq \lambda e\} \tag{1.3}$$

satisfying

$$-\|a\| e \leqq a \leqq \|a\| e . \tag{1.4}$$

Proof. Define for every $a \in A$:

$$m(a) = \inf\{\alpha \in \mathbb{R} \mid a \leqq \alpha e\}, \quad l(a) = \sup\{\beta \in \mathbb{R} \mid \beta e \leqq a\} . \tag{1.5}$$

It is easily verified that m and $-l$ are sublinear functionals. By definition $\|a\| = \max(m(a), -l(a))$ and so $\|a\|$ is a semi-norm. It is a norm by virtue of formula (1.4) which follows from Archimedicity as shown below.

For every natural number n we shall have $a \leqq [m(a) + 1/n]e$, and so

$$n[a - m(a)e] \leqq e \quad \text{for } n = 1, 2, \ldots .$$

By (1.2) this implies $a - m(a)e \leqq 0$, and a similar argument yields $l(a)e - a \leqq 0$. Hence

$$l(a)e \leqq a \leqq m(a)e,$$

from which (1.4) follows. □

We shall use the term *order-unit space*, and the notation (A, e) to denote an Archimedean ordered vector space with distinguished order unit e considered as a normed vector space in the *order-unit norm* defined by (1.3).

Proposition II.1.3. *Let $\psi : (A, e) \to (A', e')$ be a linear map between order-unit spaces, such that $\psi(e) = e'$. Now ψ is positive (order preserving) iff ψ is bounded with $\|\psi\| = 1$.*

Proof. 1. Assume first that ψ is positive and consider an element $a \in A$ such that $\|a\| \leqq 1$. Now $-e \leqq a \leqq e$, and so $-e' \leqq \psi(a) \leqq e'$. This proves $\|\psi(a)\| \leqq 1$, and so $\|\psi\| \leqq 1$. Clearly $\|\psi\| = 1$, since $\|\psi(e)\| = \|e'\| = 1$ and $\|e\| = 1$.

2. Assume $\|\psi\| = 1$ and consider a positive element $a \in A$. Without lack of generality we assume $\|a\| \leqq 1$, i.e. $0 \leqq a \leqq e$. Now $0 \leqq e - a \leqq e$, and so $\|e - a\| \leqq 1$. By assumption $\|\psi(e - a)\| \leqq 1$, and so $\psi(e - a) \leqq e'$. Hence $\psi(a) \geqq 0$, and the proof is complete. □

It follows from Proposition II.1.3 that the order and the norm of an order-unit space mutually determine each other. Specifically we shall have the following:

Corollary II.1.4. *Let $\psi : (A, e) \to (A', e')$ be a bijective linear map between order-unit spaces, such that $\psi(e) = e'$. Now ψ is an order isomorphism (ψ and ψ^{-1} both positive) iff ψ is an isometry.*

Proof. Necessity is obvious since the norms are defined by means of order and distinguished units.

Sufficiency follows from Proposition II.1.3, since an isometry ψ has the property $\|\psi\|=\|\psi^{-1}\|=1$. □

Corollary II.1.5. *A non-zero linear functional q on an order unit space (A,e) is positive iff it is bounded with $\|q\|=q(e)$.*

Proof. Apply Proposition II.1.3 to $p=q(e)^{-1}q$. □

A linear map $\psi:(A,e)\to(A',e')$ between order-unit spaces is said to be an *order homomorphism* if it satisfies the following three requirements

$$\psi(e)=e', \tag{1.6}$$

$$\psi(A^+)=\psi(A)^+, \tag{1.7}$$

$$J=J^+-J^+ \quad \text{where } J=\psi^{-1}(0). \tag{1.8}$$

Note that (1.7) is more than just positivity of ψ, which in fact is equivalent to $\psi(A^+)\subset\psi(A)^+$. The requirement (1.8) that the kernel of ψ be *positively generated* will be important in subsequent applications.

By Proposition II.1.3 every order homomorphism ψ is continuous and $\|\psi\|=1$. If $\psi:(A,e)\to(A',e')$ is a bijective order homomorphism, then ψ^{-1} is also an order homomorphism. In this case we shall say that ψ is an *isomorphism of* (A,e) *onto* (A',e'). Observe that if (A,e) is an order unit space and if e is contained in a vector subspace A_1 of A, then (A_1,e) is also an order-unit space in the induced ordering and norm. Hence we may define $\psi:(A,e)\to(A',e')$ to be an *isomorphism of* (A,e) *into* (A',e') if ψ is an isomorphism of (A,e) onto $(\psi(A),e')$. Note that $\psi:(A,e)\to(A',e')$ is an order isomorphism iff $\psi(e)=e'$, ψ is injective, and ψ and ψ^{-1} are both positive (cf. Corollary II.1.4).

If J is an order ideal of an order-unit space (A,e) and if $\varphi:A\to A/J$ is the canonical map, then $(A/J,\varphi(e))$ is *not* an order-unit space in general, since A/J may fail to be Archimedean. However, we do have the following:

Proposition II.1.6. *If J is a positively generated order ideal of an order-unit space (A,e) and if A/J is Archimedean, then $(A/J,\varphi(e))$ is an order unit space, the quotient map $\varphi:(A,e)\to(A/J,\varphi(e))$ is an order homomorphism, and the order-unit norm $\|\varphi(a)\|_0$ and the quotient norm $\|\varphi(a)\|_q$ on A/J are related as follows:*

$$\|\varphi(a)\|_0\leqq\|\varphi(a)\|_q, \quad \text{all } a\in A.$$

In particular J is a closed subspace of A.

Proof. The first part of the proposition is easily verified. To prove (1.9) we consider an arbitrary element $a \in A$. Now

$$-\|a+b\|e \leqq a+b \leqq \|a+b\|e, \quad \text{all } b \in J.$$

Hence

$$-\|a+b\|\varphi(e) \leqq \varphi(a) \leqq \|a+b\|\varphi(e), \quad \text{all } b \in J,$$

and so

$$\|\varphi(a)\|_0 \leqq \|a+b\|, \quad \text{all } b \in J.$$

Now it follows that $\|\varphi(a)\|_0 \leqq \|\varphi(a)\|_q$. □

Remark. The two norms of (1.9) are not equal in general. As an example one may consider the space $A = A(K)$ of all (continuous) affine functions on the trapezoid K shown in the diagram, with $F : F' = \lambda < 1$.

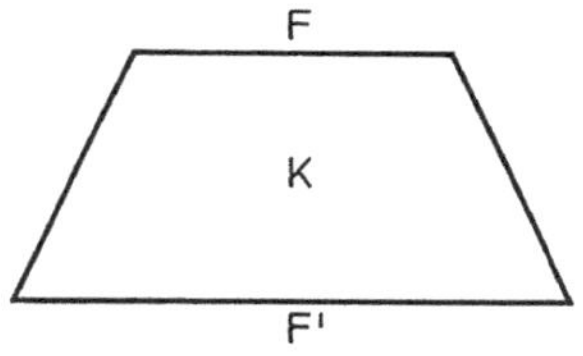

Fig. 2

Clearly $(A, 1_K)$ is an order-unit space in standard (pointwise) ordering and uniform norm. Also it is easily seen that $J = \{a \in K \mid a(F) = 0\}$ is an order ideal for which $(A/J, \varphi(1_K))$ is an order-unit space isomorphic to $(A(F), 1_F)$. Interpreting the two norms we see that $\|-\|_0$ is the uniform norm over F, while $\|-\|_0$ is the infimum of the uniform norms of all possible affine *extensions* to K of a given affine function on F. Hence

$$\lambda\|\varphi(a)\|_q \leqq \|\varphi(a)\|_0 \leqq \|\varphi(a)\|_q, \quad \text{all } a \in A, \tag{1.9}$$

and this relation can not prevail with any larger constant in the place fo λ.

Let $\psi : (A, e) \to (A', e')$ be some order-homomorphism between order-unit spaces, let $J = \psi^{-1}(0)$ be the kernel of ψ and let φ be the canonical map of A onto A/J. Now it is easily seen that $(A/J, \varphi(e))$ is an order unit space, that φ is an order homomorphism, and that ψ can be decomposed as $\psi=\psi'\circ\varphi$, where ψ' is an isomorphism of $(A/J,\varphi(e))$ into (A',e').

We shall say that an order homomorphism $\psi:(A,e)\to(A',e')$ is a *topological order homomorphism* if it is an open mapping of A onto $\psi(A)$ (in other words, if ψ is a "strict morphism" of topological vector spaces [85]). Decomposing $\psi=\psi'\circ\varphi$, we observe that φ is an open mapping of A onto A/J provided with the quotient norm, while ψ' becomes an isometry when A/J is provided with the order-unit norm determined by $\varphi(e)$ (Corollary II.1.4.) Hence *ψ is a topological order homomorphism iff the two norms on A/J are topologically equivalent*, i.e. if there exists a strictly positive constant λ such that a relation (1.9) is valid.

A linear functional p on an order-unit space (A,e) is called a *state* if $p\geqq 0$ and $p(e)=1$. By Corollary II.1.5 this is equivalent to

$$\|p\|=p(e)=1\,. \tag{1.10}$$

The states of (A,e) form a w^*-compact convex subset of A^* which we shall call the *state space* of (A,e) and denote by $S(A,e)$ or simply by S.

The crucial property of order-unit spaces is the existence of "sufficiently many" states.

Proposition II.1.7. *If a is an element of an order-unit space (A,e), then*

$$a\geqq 0 \;\Leftrightarrow\; p(a)\geqq 0 \quad \text{all } p\in S\,, \tag{1.11}$$

and

$$\|a\|=\sup\{|p(a)|\,|\,p\in S\}\,. \tag{1.12}$$

Moreover, the unit ball of A^* is given by

$$A_1^*=\operatorname{conv}(S\cup -S)\,. \tag{1.13}$$

Proof. 1. If $a\geqq 0$, then by definition $p(a)\geqq 0$ for all $p\in S$.

Conversely we assume $a\notin A^+$. Now $l(a)<0$ where l is the functional defined in (1.5), and it is easily verified that

$$\{b\in A\,|\,\|a-b\|<\tfrac{1}{2}|l(a)|\}\subset \complement A^+\,.$$

Now we may apply the Hahn-Banach Theorem to separate a from A^+. Thus there is an $f\in A^*$ such that $f(a)<\alpha\leqq f(A^+)$ for some $\alpha\in\mathbb{R}$. Since A^+ is a cone, we can take $\alpha=0$. Now $p=f(e)^{-1}f$ will be a state with $p(a)<0$, and the first part of the proof is complete.

2. Clearly $\|a\|\geqq\sup\{|p(a)|\,|\,p\in S\}$. To prove the converse relation we assume $\|a\|=m(a)$, where m is the functional defined in (1.5). This

can be done without loss of generality, for if we had $\|a\| = -l(a)$ we should simply replace a by $-a$.

By definition

$$(m(a)-\varepsilon)e-a \notin A^+$$

for every $\varepsilon>0$.

By the first part of the proposition, there is a $p_0 \in S$ such that

$$p_0[(m(a)-\varepsilon)e-a]<0\,.$$

Hence

$$m(a)-\varepsilon<p_0(a)\,,$$

and since $\varepsilon>0$ was arbitrary:

$$m(a) \leqq \sup\{p(a) \mid p \in S\}\,.$$

This completes the second part of the proof.

3. Clearly $\operatorname{conv}(S \cup -S) \subset A_1^*$. To prove the converse relation, we consider an element $q \notin \operatorname{conv}(S \cup -S)$. By the w^*-compactness of S the set $\operatorname{conv}(S \cup -S)$ is also w^*-compact, hence closed. Hence there exists a w^*-continuous linear functional on A^*, i.e. an element a of A, which separates q strictly from $\operatorname{conv}(S \cup -S)$. Since $\operatorname{conv}(S \cup -S)$ is a symmetric set, we may take the separating real number to be 1. Thus

$$q(a)>1, \quad p(a) \leqq 1 \quad \text{for } p \in \operatorname{conv}(S \cup -S)\,.$$

The last inequality means that $|p(a)| \leqq 1$ for all $p \in S$. By the second part of the proposition this is equivalent to $\|a\| \leqq 1$. Hence $q \notin A_1^*$ and the proof is complete. □

Let (A,e) be an order unit space and X some compact Hausdorff space. By Corollary II.1.4 a linear map $\rho:(A,e) \to (C_{\mathbb{R}}(X), 1_X)$ is an *isomorphism* (into) iff it satisfies the condition (1.14) and the two mutually equivalent conditions (1.15) and (1.16):

$$\rho(e)=1_X, \tag{1.14}$$

$$\rho \text{ is } 1-1 \text{ and both ways order preserving}, \tag{1.15}$$

$$\rho \text{ is an isometry.} \tag{1.16}$$

An isomorphism $\rho:(A,e) \to (C_{\mathbb{R}}(X), 1_X)$ will be called a *functional representation of* (A,e) *over* X, and it will be denoted by the symbol (ρ, X). A functional representation (ρ, X) will be said to be *separating* if $\rho(A)$ separates the points of X.

Let (ρ, X) and (σ, Y) be two functional representations of (A, e). We shall say that (ρ, X) is *larger* than (σ, Y) if there exists a homeomorphism $\varphi: Y \to X$ (into) such that $\sigma = \varphi^* \circ \rho$, or specifically

$$[\sigma a](y) = [\rho a](\varphi y), \quad \text{all } a \in A,\ y \in Y. \tag{1.17}$$

Note that if (ρ, X) is a functional representation of (A, e), and if a closed subset Y of X is a *max-boundary* for $\rho(A)$ in that

$$\|a\| = \|\rho a\| = \sup_{y \in Y} |[\rho a](y)|, \quad \text{all } a \in A, \tag{1.18}$$

then the canonical injection $\varphi: Y \to X$ induces a functional representation $(\varphi^* \circ \rho, Y)$ which we term the *restriction* of (ρ, X) to Y.

Observe that by Proposition II.1.7 every order-unit space (A, e) really admits a functional representation. One such is the *canonical representation* (ρ, S) *over the state space*, defined by $a \rightsquigarrow \tilde{a}$ where $\tilde{a}(p) = p(a)$ for all $p \in S$.

Theorem II.1.8. *(Kadison) The canonical representation (ρ, S) of an order-unit space (A, e) over its state space S is the largest separating functional representation of (A, e). Its range $\rho(A)$ consists of all those w^*-continuous affine functions on S which can be extended to w^*-continuous linear functionals on the surrounding space A^*. In particular, $\rho(A) = A(S)$ iff (A, e) is complete in order-unit norm. The restriction of (ρ, S) to $\overline{\partial_e S}$ is the smallest separating functional representation of (A, e).*

Proof. 1. By the definition of ρ its range $\rho(A)$ consists of all restrictions to S of evaluation functionals $f \rightsquigarrow f(a)$ where $a \in A$ and f runs through A^*, and these are exactly the w^*-continuous linear functionals on A^*. The set S is located on the w^*-closed hyperplane $H = \{f \in A^* \mid f(e) = 1\}$, and it follows that every w^*-continuous affine function on A^* coincides on S with some w^*-continuous linear functional on A^* (redefine it at $0 \notin H$). Hence $\rho(A) = A(S, A^*)$, where A^* is endowed with the w^*-topology. By Corollary I.1.5 $A(S, A^*)$ is uniformly dense in $A(S)$, and by the isometric nature of ρ the two spaces will coincide iff A is complete.

2. To prove the maximality of (ρ, S) we consider an arbitrary separating functional representation (σ, Y). To every point $y \in Y$ we associate a state φy defined by

$$\varphi y : a \rightsquigarrow [\sigma a](y). \tag{1.19}$$

The map $\varphi: Y \to S$ is continuous since the functions σa are continuous on Y for every $a \in A$, and it is injective since $\sigma(A)$ separates the points of Y. Hence φ is a homeomorphism of Y into S. Also it follows from the definition (1.19) that $\sigma = \varphi^* \circ \rho$.

3. Every continuous affine function on S assumes its maximum modulus on $\partial_e S$ (Krein-Milman). Hence $\overline{\partial_e S}$ is a max-boundary for $A(S,A^*)=\rho(A)$, and the restriction of (ρ,S) to $\overline{\partial_e S}$ is well defined.

To prove the minimality of this restriction we consider an arbitrary separating functional representation (σ,Y). We define φ as above, and consider an arbitrary $a\in A$ with canonical representative $\rho a=\tilde{a}$ over S. Since all occurring representations are isometries, we shall have

$$\Big|\sup_{y\in Y}\tilde{a}(\varphi y)\Big|=\sup_{y\in Y}|[\sigma a](y)|=\|a\|=\sup_{p\in S}|\tilde{a}(p)|. \tag{1.20}$$

Hence $\varphi(Y)$ is a max-boundary for $\rho(A)=A(S,A^*)$. Now it follows by Milman's Theorem that $\varphi(Y)\supset\partial_e S$. In fact if $p\in\partial_e S\backslash\varphi(Y)$, then $p\notin \mathrm{cl.conv.}\,\varphi(Y)$, and then p could be separated from $\varphi(Y)$ by some w^*-continuous affine function on A^* assuming its max-value over S on a subset disjoint from $\varphi(Y)$.

Now $\overline{\partial_e S}\subset\varphi(Y)$, and the canonical injection $\pi:\overline{\partial_e S}\to S$ may be decomposed as $\pi=\varphi\circ\tau$ where τ is a homeomorphism of $\overline{\partial_e S}$ into Y. Dualizing, we get $\pi^*=\tau^*\circ\varphi^*$, and so $\pi^*\circ\rho=\tau^*\circ\varphi^*\circ\rho=\tau^*\circ\sigma$. Hence $(\pi^*\circ\rho,\overline{\partial_e S})$ is a smaller functional representation than (σ,Y), and the proof is complete. □

Proposition II.1.9. *Let (A,e) be an order-unit space which is a lattice in the given ordering. A linear functional p on A belongs to $\partial_e S$ iff p is lattice preserving and $p(e)=1$.*

Proof. Let $a,b\in A$ and let $\tilde{a},\tilde{b}$ be their (canonical) representatives over S. By the definition of upper envelope (Ch. I, (1.4)), we shall have

$$\begin{aligned}\widehat{(\tilde{a}\vee\tilde{b})}(p)&=\inf\{\tilde{c}(p)\,|\,\tilde{a},\tilde{b}\leqq\tilde{c},\ c\in A\}\\ &=\inf\{p(c)\,|\,a,b\leqq c,\ c\in A\}=p(a\vee b).\end{aligned} \tag{1.21}$$

Hence a state p is lattice preserving iff

$$\widehat{(\tilde{a}\vee\tilde{b})}(p)=(\tilde{a}\vee\tilde{b})(p),\quad \text{all } a,b\in A\,. \tag{1.22}$$

It follows from Hervé's result (Prop. I.4.1) that (1.22) is valid for every $p\in\partial_e S$.

Conversely, if p is a lattice preserving linear functional and $p(e)=1$, then p is a state, and by virtue of (1.22) $\hat{f}(p)=f(p)$ for all functions $f=\tilde{a}_1\vee\cdots\vee\tilde{a}_n$, where $a_i\in A$ for $i=1,\dots,n$. The set of all such functions is a dense subcone $P_0(S)$ of $P(S)$ (Cor. I.1.3), and a slight modification of the proof of Proposition I.4.1 shows that

$$\hat{f}(p)=f(p)\quad \text{all } f\in P_0(S)\ \Rightarrow\ p\in\partial_e S\,. \tag{1.23}$$

This completes the proof. □

Theorem II.1.10. *(Stone-Kakutani-Krein-Yosida) If an order-unit space (A,e) is a lattice in the given ordering, then $\partial_e S$ is w*-closed, and the (minimal) functional representation of (A,e) over $\overline{\partial_e S}=\partial_e S$ is a lattice isomorphism of A onto a dense lattice-subspace of $C_{\mathbb{R}}(\partial_e S)$.*

Proof. It is easily verfied that the set of lattice preserving linear functionals is w^*-closed in S, hence $\overline{\partial_e S}=\partial_e S$. It also follows from Proposition II.1.9 that the representation over $\partial_e S$ is a lattice isomorphism, and the density in $C_{\mathbb{R}}(\partial_e S)$ follows by the lattice version of the Stone-Weierstrass Theorem. □

Corollary II.1.11. *An order-unit space (A,e) is isomorphic to $C_{\mathbb{R}}(X)$ for some compact Hausdorff space X iff A is a lattice in the given ordering and A is complete in order-unit norm.*

We shall now pass to the study of another class of ordered vector spaces, and we shall show that there is a natural duality between those and the order-unit spaces studied so far. Recall that an ordered vector space E is *directed* if $E=E^+-E^+$, and that a convex subset B of a vector space E is *radially compact* if $B\cap L$ is a closed and bounded segment for every line L through the origin of E. Recall also that a convex subset K of a hyperplane H not passing through the origin of a vector space E, is said to be a *base* for the cone $C=\bigcup_{\lambda\geqq 0}\lambda K$.

Proposition II.1.12. *If E is a directed vector space for which the cone E^+ has a base K such that $B=conv(K\cup -K)$ is radially compact, then E is organized to a normed space by the Minkowski functional*

$$\|x\|=\inf\{\lambda\geqq 0\,|\,x\in\lambda B\}\,, \tag{1.24}$$

and the closed unit ball is equal to B.

If K is compact in some Hausdorff topology $\mathcal{T}$ compatible with the linear structure of E, then E is complete in the norm (1.24).

Proof. The set B is absorbing since E is directed. Also B is balanced and convex. Hence the norm is well defined and the stated properties are easily verified.

Assuming that K and hence B is $\mathcal{T}$-compact, we consider a (norm) Cauchy sequence $\{x_n\}$. Now $\{x_n\}$ is bounded, and so we may assume $x_n\in B$ for $n=1,2,\ldots$. Let y be some $\mathcal{T}$-accumulation point of $\{x_n\}$ in B. We claim that y must be a norm limit of $\{x_n\}$.

Let $\varepsilon>0$ be arbitrary and choose a natural number n_0 such that $\|x_n-x_m\|\leqq\varepsilon$ for $n,m\geqq n_0$. In particular $x_n\in x_{n_0}+\varepsilon B$ for $n\geqq n_0$. Since εB is $\mathcal{T}$-closed, we shall have $y\in x_{n_0}+\varepsilon B$, or equivalently $\|y-x_{n_0}\|\leqq\varepsilon$. Hence

$$\|y-x_n\|\leqq\|y-x_{n_0}\|+\|x_{n_0}-x_n\|\leqq 2\varepsilon,$$

for all $n\geqq n_0$, and the proof is complete. □

We shall use the term *base-norm space* and the notation (E,K) to denote a directed vector space E for which E^+ has a base K such that $B=\operatorname{conv}(K\cup -K)$ is radially compact, considered as a normed space in the *base-norm* (1.24).

If (E,K) is a base norm space, then by definition there is a linear functional f on E such that $K\subset f^{-1}(1)$. By the directedness f is uniquely determined. We shall call it the linear functional which *carries* K, and we shall henceforth denote it by the symbol e_K. It is easily verified that e_K is a bounded linear functional on (E,K).

Proposition II.1.13. *The norm of a base-norm space* (E,K) *is additive on* E^+. *In fact* $\|x\|=e_K(x)$ *for* $x\in E^+$.

Proof. If $x\in E^+$ then $x=\rho x_0$, where $x_0\in K$ and $\rho\geqq 0$. Now $x\in\rho B$, and so $\|x\|\leqq\rho$. Applying e_K to $x=\rho x_0$, we get $e_K(x)=\rho$. Hence we get $\|x\|\leqq e_K(x)$.

Conversely $x\in\|x\|B$, and so there are elements $y,z\in K$ and psitive numbers λ,μ such that

$$x=\|x\|(\lambda y-\mu z),\qquad \lambda+\mu=1.$$

Applying e_K to this equation we obtain $e_K(x)=\|x\|(\lambda-\mu)\leqq\|x\|$, and the proof is complete. □

Proposition II.1.14. *Every element* x *of a base-norm space* (E,K) *admits a decomposition* $x=y-z$, *where* $y,z\geqq 0$ *and* $\|x\|=\|y\|+\|z\|$.

Proof. It follows from the relation $x\in\|x\|B$ that there exist elements $y_0,z_0\in K$ and positive numbers λ_0,μ_0 such that

$$x=\lambda_0\|x\|y_0-\mu_0\|x\|z_0,\qquad \lambda_0+\mu_0=1.$$

Writing $y=\lambda_0\|x\|y_0$ and $z=\mu_0\|x\|z_0$ we obtain $x=y-z$ and

$$\|y\|+\|z\|=(\lambda_0+\mu_0)\|x\|=\|x\|.$$

This completes the proof. □

Theorem II.1.15. *(Ellis) If (A,e) is an order-unit space, then (A^*,S) is a base-norm space, and the base-norm is the usual norm of A^* considered as the dual of A with the order-unit norm. Conversely, if (E,K) is a base-norm space, then (E^*,e_K) is an order-unit space, and the order-unit norm is the usual norm of E^* considered as the dual of E with the base-norm.*

Proof. 1. Clearly A^* is an ordered vector space in the usual ordering, i.e. $q \geqq 0$ iff $q(a) \geqq 0$ for all $a \in A^+$. Also it is seen that S is a base of $(A^*)^+$, and it follows from Proposition II.1.7 that $\operatorname{conv}(S \cup -S)$ is the closed unit ball of the normed space A^*. Hence (A^*,S) is a base-norm space, and the base-norm is the usual norm of A^*.

2. Clearly E^* is an Archimedean ordered vector space in the usual ordering, i.e. $f \geqq 0$ iff $f(x) \geqq 0$ for all $x \in E^+$. Also it is clear that e_K is a continuous positive linear functional on E.

For every $f \in E^*$ we shall have

$$\begin{aligned}
\|f\| &= \sup\{|f(x)| \,|\, x \in B\} \\
&= \sup\{|f(x)| \,|\, x \in K\} \\
&= \inf\{\lambda \geqq 0 \,|\, -\lambda \leqq f(x) \leqq \lambda, \text{ all } x \in K\} \\
&= \inf\{\lambda \geqq 0 \,|\, -\lambda e_K(x) \leqq f(x) \leqq \lambda e_K(x), \text{ all } x \in K\} \\
&= \inf\{\lambda \geqq 0 \,|\, -\lambda e_K \leqq f \leqq \lambda e_K\}.
\end{aligned}$$

Hence e_K is an order-unit of E^*, and the order-unit norm is the usual norm of E^*. This completes the proof. □

Remark. Theorem II.1.15 may be stated more briefly as follows: The dual of an order-unit space (A,e) is the base-norm space (A^*,S), and the dual of a base-norm space (E,K) is the order-unit space (E^*,e_K).

Notes. Theorem II.1.8 is due to Kadison. In 1951 he showed that the canonical representation (ρ,S) of an (Archimedean) order-unit space (A,e), as well as the restriction of ρ to $\overline{\partial_e S}$, are "functional representations", i.e. he showed them to be linear order-isomorphisms and isometries mapping e onto the constant 1 [224]. The equality of $\rho(A)$ and $A(S)$ for a complete A was proved by Kadison in 1963 [228] (Lem. 4.3), and by an independent and somewhat simpler argument by Semadeni in 1965 [361]. The maximality and minimality properties were noted by Kadison in 1957 [226]. Theorem II.1.10 is the well known vector lattice representation theorem. It dates back to 1940–41, and it is due to Stone [376], Kakutani [235], Krein [252], and Yosida [405]. Another relevant references is Nachbin's 1949-paper [305]. Our formulation of the theorem is essentially that of Yosida, but the proof is based on that of Kadison [224]. In this connection we should like to point out that Hervé's characterization of extreme points is essential in this proof, and that Hervé's result in fact is implicit in Kadison's proof [224]. The theory of partially ordered linear spaces, and in particular of order-unit spaces and their duals, has been studied by a group of British mathematicians, and in particular by Bonsall [77], [78], [79], [80], Edwards [142], and Ellis [164]. Theorem II.1.15 of the preceding section was proved by Ellis in 1964 [164].

§ 2. Elementary Embedding Theorems

Let X be an arbitrary compact Hausdorff space. We shall say that a vector subspace of $C_{\mathbb{R}}(X)$ is an *admissible space* (over X) if it contains the constant functions and separates points. If A is an admissible space, then A is also an admissible cone (Ch. I, § 5), and $(A, 1_X)$ is an order-unit space. To every $x \in X$ is associated a state $\tilde{x}$ on $(A, 1_X)$, the *point state* at x, defined by $\tilde{x}(a) = a(x)$ for all $a \in A$. Observe also that the restriction map $\psi: C_{\mathbb{R}}(X)^* \to A^*$ maps $M_1^*(X)$ *onto* S, since every $p \in S$ can be extended by the Hahn-Banach Theorem to a measure μ such that $\|\mu\| = \mu(1_X) = 1$. The set $\psi^{-1}(p) \cap M_1^+(X)$ will be denoted by the symbol $M_{p,A}^+(X)$. It consists of all measures in $M_1^+(X)$ which *represent* the state p, in that

$$p(a) = \int a \, d\mu \quad \text{all } a \in A. \tag{2.1}$$

For every $x \in X$ we shall write $M_{x,A}^+(X)$ in the place of $M_{\tilde{x},A}^+(X)$, and we note that this conforms with previous notation (Ch. I, § 5).

Theorem II.2.1. *If A is an admissible space over a compact Hausdorff space X, then $\varphi: x \to \tilde{x}$ is a homeomorphic embedding of X into the state space S of $(A, 1_X)$, and it is related to the canonical representation $\rho: a \to \tilde{a}$ of $(A, 1_X)$ over S by the formula*

$$\tilde{a}(\tilde{x}) = a(x), \quad \text{all } x \in X, \quad a \in A. \tag{2.2}$$

Moreover, φ maps the Choquet boundary $\partial_A X$ onto the extreme boundary $\partial_e S$.

Proof. 1. The map φ is continuous since S is endowed with the w^*-topology, and it is $1-1$ since A separates points. Hence φ is a homeomorphism of X into S. By definition we have $\tilde{a}(\tilde{x}) = \tilde{x}(a) = a(x)$ for all $x \in X$ and all $a \in A$.

2. To prove $\varphi(\partial_A X) \subset \partial_e S$, we assume this relation inexact, say $x \in \partial_A X$ and $\tilde{x} \notin \partial_e S$. Then there exist $p, q \in S$ such that $x = \frac{1}{2}p + \frac{1}{2}q$ and $p \neq q$. Now let $\mu \in M^+_{p,A}(X)$ and $\nu \in M^+_{q,A}(X)$. Clearly

$$\tfrac{1}{2}\mu + \tfrac{1}{2}\nu \in M^+_{x,A}(X).$$

Since x is a Choquet-point, this implies

$$\tfrac{1}{2}\mu + \tfrac{1}{2}\nu = \varepsilon_x.$$

This, however, is impossible since μ and ν are distinct probability measures.

3. To prove $\partial_e S \subset \varphi(\partial_A X)$ we consider an arbitrary $p \in \partial_e S$, and we shall prove that $M^+_{p,A}(X) = \{\varepsilon_x\}$ for some $x \in X$, this being equivalent with $p = \tilde{x}$, $x \in \partial_A X$.

The set $M^+_{p,A}(X) = \psi^{-1}(p) \cap M^+_1(X)$ is a (vaguely-) closed face of $M^+_1(X)$. It is the closed convex hull of its extreme points; these are extreme in $M^+_1(X)$ as well, hence of the form ε_x for $x \in X$ (cf. e.g. [140]). But the set $M^+_{p,A}(X)$ can not contain more than one such measure, since A separates points. This completes the proof. □

By Theorem II.2.1. the embedding $\varphi: X \to S$ carries the functions $a \in A$ into functions extendable (with preservation of norm) to continuous and *affine* functions a on S. In fact, all functions in $A(S, A^*_w)$ are obtained in this way, and $A(S, A^*_w) = A(S)$ iff A is uniformly closed in $C_{\mathbb{R}}(X)$ (Theorem II.1.8. The subscript w indicates that A^* is endowed with the w^*-topology.). Note also that for admissible spaces one can obtain the boundary measure representations of Ch. I, § 5 from the corresponding geometric theory by means of the embedding $\varphi: X \to S$.

Assume for the moment that K is a compact convex subset of a locally convex Hausdorff space E and that the following condition is satisfied:

$$E = \operatorname{lin}(K) \quad \text{and} \quad K \subset H, \quad \text{where } H \text{ is a hyperplane such that } 0 \notin H. \tag{2.3}$$

From (2.3) it follows easily that H is equal to the affine span of K. Also it is easily seen that every point $x \in E$ defines a unique linear (but not necessarily continuous) functional q_x on $A(K)$ such that

$$q_x = \lambda a(y) - \mu a(z), \quad \text{all } a \in A(K), \tag{2.4}$$

whenever

$$x = \lambda y - \mu z; \quad y, z \in K; \quad \lambda, \mu \in \mathbb{R}^+.$$

Clearly the map $x \rightsquigarrow q_x$ is linear.

We shall say that a compact convex set K is *regularly embedded* in a locally convex Hausdorff space E if (2.3) is satisfied, and if $x \rightsquigarrow q_x$ is a topological isomorphism of E onto the space $A(K)^*_w$, i.e. onto the norm dual of $A(K)$ endowed with the w^*-topolgy.

Proposition II.2.2. *If K is a compact convex set which is regularly embedded in a locally convex Hausdorff space E, then:*

(i) *Every continuous affine function on K can be extended (uniqely) to a continuous linear functional on E.*

(ii) *The hyperplane H containing K is closed.*

(iii) *A convex subset G of E is closed iff the sets $(1/n)G \cap \operatorname{conv}(K \cup -K)$ are closed for $n=1,2,\dots$.*

Proof. 1. For a given $a \in A(K)$ the function $x \rightsquigarrow q_x(a)$ is a continuous linear functional on E, and it extends a since $q_x(a)=a(x)$ for $x \in K$.

2. The second statement follows from the first, since the constant 1_K has a continuous linear extension f to E, and $H=\{x \in E \mid f(x)=1\}$.

3. The third statement follows from the Krein-Šmulian Theorem (Cor. I.1.13), since $x \rightsquigarrow q_x$ is a topological isomorphism of E onto a Banach dual space endowed with the w^*-topology and this isomorphism maps $\operatorname{conv}(K \cup -K)$ onto the closed unit ball. □

Corollary II.2.3. *If a compact convex set K is regularly embedded in a locally convex vector space E, then $A(K;E)=A(K)$.*

It is of interest to observe that every compact convex subset K of a locally convex (Hausdorff) space E_0 can be "regularly embedded" in some (other) locally convex Hausdorff space E. In fact we shall state this result in a form which dispenses altogether with the initial space E_0, and in this connection we shall need a general definition:

An *abstract compact convex* is a pair (X,A) where X is a compact Hausdorff space and A is a uniformly closed, admissible subspace of $C_{\mathbb{R}}(X)$ such that every state on $(A, 1_X)$ is a point state.

If K is a compact convex subset of a locally convex Hausdorff space, then $(K, A(K))$ is an abstract compact convex in the above sense, since an arbitrary state p on $(A(K), 1_K)$ is a point state $\tilde{x}$, x being the barycenter of any $\mu \in M_1^+(K)$ representing p.

A *regular embedding* of an abstract compact convex (X,A) into a locally convex Hausdorff space E is a homeomorphism φ of X into E such that $\varphi(X)$ is a compact convex set regularly embedded in E and the map φ^* defined by

$$[\varphi^* a](x)=a(\varphi(x)), \qquad \text{for all } x \in X, \quad a \in A(\varphi(X)), \tag{2.5}$$

is an isomorphism of $A(\varphi(X))$ onto A.

If (X,A) is of the type $(K,A(K))$, i.e. if (X,A) is a "concrete" compact convex, then φ must be *affine* by virtue of (2.5), and in this case we shall briefly say that φ is a regular embedding of K rather than of $(K,A(K))$.

Theorem II.2.4. *Every abstract compact convex (X,A) admits a regular embedding into a locally convex Hausdorff space. In fact the map $\varphi: X \to A_w^*$ defined by $x \rightsquigarrow \tilde{x}$ is such an embedding, and it maps X onto the state space S of $(A, 1_X)$. Moreover, this regular embedding is unique, in that for every other regular embedding $\varphi': X \to E'$ there exists a linear homeomorphism η of E' onto A_w^* such that $\varphi = \eta \circ \varphi'$, $\varphi' = \eta^{-1} \circ \varphi$.*

Proof. By Theorem II.2.1 $\varphi: x \rightsquigarrow \tilde{x}$ is a homeomorphism of X into S. It is *onto* since every state on $(A, 1_X)$ is a point state. Clearly S is regularly embedded in A_w^*, and the requirement (2.5) is satisfied by virtue of (2.2). Hence we have proved that $\varphi: X \to A_w^*$ is a regular embedding. To prove uniqueness, it suffices to consider the map $\eta: E' \to A(S)_w^* \cong A_w^*$ which sends every $x \in E'$ into the linear functional q_x defined by (2.4), and to use the definitions. □

Henceforth we shall refer to the map φ of Theorem II.2.4 as *the regular embedding* of the given compact convex. We note that there is no more need to distinguish between "abstract" and "concrete" compact convexes; nor is there any need to specify a given surrounding space E_0 of a ("concrete") compact convex, unless one wants to study non-intrinsic properties (such as e.g. the existence of supporting hyperplanes in E_0).

If K is a compact convex set and $\varphi: K \to E = A(K)_w^*$ is the regular embedding of K, then we define the *associated cone* $\tilde{K}$ to be the cone in E with summit 0 and base $\varphi(K)$. Observe that we can identify $\tilde{K}$ with the cone of positive elements of $A(K)^*$ endowed with the usual ordering.

Many properties of K, e.g. its facial structure, can be most conveniently studied by means of the associated cone $\tilde{K}$.

Generally a subcone C' of a cone C is said to be *hereditary* (downwards) if

$$x \in C', \quad y \in C \cap (x - C) \Rightarrow y \in C'. \tag{2.6}$$

Now we have the following:

Proposition II.2.5. *A convex subset F of a compact convex set K is a face iff $\tilde{F}$ is an hereditary subcone of $\tilde{K}$.*

Proof. For simplicity we assume that K is already regularly embedded in E; in particular we assume $K \subset H$ where $H = \{x \in E \mid f(x) = 1\}$ and $f \in E^*$.

1. Assume first that F is a face of K and consider elements $x \in \tilde{F}$, $y \in \tilde{K} \cap (x - \tilde{K})$. If $y = 0$ or $y = x$, then there is nothing to prove. Otherwise $0 < f(y) < f(x)$, and hence also $0 < f(z) < f(x)$, where $z = x - y$. Writing $x' = f(x)^{-1} x$, $y' = f(y)^{-1} y$, $z' = f(z)^{-1} z$, we obtain $x' \in F$ and $y', z' \in K$. Also we can express x' as the following convex combination of y' and z'

$$x' = \frac{f(y)}{f(x)} y' + \frac{f(z)}{f(x)} z' . \tag{2.7}$$

Since F is a face of K we shall have $y' \in F$, and hence $y = f(y) y' \in \tilde{F}$.

2. Assume next that $\tilde{F}$ is an hereditary subcone and consider a convex combination $x = \lambda y + (1 - \lambda) z$ where $x \in F$ and $y, z \in K$ and where $0 < \lambda \leqq 1$. Now $\lambda y \in \tilde{K} \cap (x - \tilde{K})$, and by assumption $\lambda y \in F$. Hence $y \in K \cap \tilde{F} = F$, and the proof is complete. □

Finally we observe that the cone $\tilde{K}$ associated with a compact convex set K is locally compact by virtue of the following:

Theorem II.2.6. *(Klee) A proper cone C in a locally convex Hausdorff space E is locally compact iff it has a compact base.*

Proof. 1. If C is locally compact, then there is a closed convex neighbourhood V of 0 in E such that $V \cap C$ is compact. Let D be the frontier of $V \cap C$ for the *relative* topology of C. Now $C = \{\lambda x \mid x \in D, \lambda \geqq 0\}$ and $0 \notin D$. The point 0 is an extreme point of $V \cap C$ since C is a proper cone, and it follows by Milman's Theorem that $0 \notin \text{cl.conv.} D$. By Hahn-Banach separation there is a closed hyperplane H separating 0 and D. Now $H \cap C$ is a base, and it is compact since it is a closed subset of the compact set $V \cap C$.

2. Next we assume that C has a compact base, i.e. we assume that there exists a hyperplane H such that $0 \notin H$ and $K = H \cap C$ is compact. Without lack of generality we can assume that H is closed. (Otherwise separate 0 and K by a closed hyperplane H' and replace K by $K' = H' \cap C$.) Let $H = \{x \in E \mid f(x) = 1\}$ where $f \in E^*$, and consider for an arbitrary $x \in C$ a real number α such that $\alpha > f(x)$. Now $V = \{y \in E \mid f(y) \leqq \alpha\}$ is a closed neighbourhood of x and $V \cap C = \text{conv}(\{0\} \cup \alpha K)$ is compact. Hence C is locally compact. □

Remark. It is not hard to prove that a locally compact proper cone C must be *closed*. However, we shall not need this result in the sequel. (For proof cf. e.g. [142].)

For later references we note the following:

Corollary II.2.7. *If a proper cone C in a locally convex Hausdorff space E is locally compact, then $C \cap (x - C)$ is compact for every $x \in C$.*

Proof. By Theorem II.2.6 C admits a compact base K, and without lack of generality we may assume that x is contained in the compact set $K' = \operatorname{conv}(\{0\} \cup K)$. Now the equality

$$C \cap (x - C) = K' \cap (x - K')$$

completes the proof. □

Notes and remarks. Most of the material of the preceding section can be regarded as mathematical folklore. A proof of Theorem II.2.6 was given in 1955 by Klee [245].

§ 3. Choquet Simplexes

Let K be a compact convex set and without loss of generality we assume K regularly embedded in E $(\cong A(K)^*_w)$. We define K to be a *Choquet simplex*, or briefly a *simplex* if E is a lattice in the ordering defined by the cone $\tilde{K}$ with base K.

If K is a simplex, then E is a vector lattice with $E^+ = \tilde{K}$, and it follows that E has the *Riesz interpolation property*:

$$x_1, x_2 \leqq y_1, y_2 \Rightarrow \exists z : x_1, x_2 \leqq z \leqq y_1, y_2, \tag{3.1}$$

which is equivalent to the *Riesz decomposition property*:

$$\begin{aligned} 0 \leqq u \leqq v_1 + v_2, \quad 0 \leqq v_1, v_2 \Rightarrow \exists u_1, u_2 : 0 \leqq u_1 \leqq v_1, \\ 0 \leqq u_2 \leqq v_2, \quad u_1 + u_2 = u, \end{aligned} \tag{3.2}$$

by virtue of the following:

Proposition II.3.1. *A (partially) ordered vector space E enjoys the Riesz interpolation property iff it enjoys the Riesz decomposition property.*

Proof. 1. Assume first (3.1) and consider elements u, v_1, v_2 such that $0 \leqq u \leqq v_1 + v_2$, $0 \leqq v_1, v_2$. Define $x_1 = u - v_1$, $x_2 = 0$, $y_1 = u$, $y_2 = v_2$, and apply (3.1) to yield an element z such that $x_1, x_2 \leqq z \leqq y_1, y_2$. Now the elements $u_1 = u - z$ and $u_2 = z$ will satisfy the requirements of (3.2).

2. Assume next (3.2) and consider elements x_i, y_j such that $x_i \leqq y_j$ for $i,j=1,2$. Define $v_1=y_1-x_1$, $v_2=y_2-x_2$, $u=y_1-x_2$, and apply (3.2) to yield elements u_i such that $0\leqq u_i \leqq v_i$ for $i=1,2$ and $u_1+u_2=u$. Now the element $z=x_2+u_2$ will satisfy the requirements of (3.1). □

An ordered vector space with the Riesz interpolation property is not a vector lattice in general. However, the following proposition is valid:

Proposition II.3.2. *If E is a directed vector space with the Riesz interpolation property and E^+ is locally compact in some locally convex Hausdorff topology, then E is a vector lattice.*

Proof. It suffices to prove that any two elements x, y of E^+ have a least upper bound in E^+. (Cf. the proof of Proposition I.1.1.) For every $z\geqq x, y$ let $K_z=\{w \mid x, y \leqq w \leqq z\}$. By Corollary II.2.7 each K_z is compact, and by the Riesz interpolation property the family of all sets K_z with $z \geqq x, y$ has the finite intersection property. It follows that

$$\bigcap_{z\geqq x,y} K_z \neq \emptyset,$$

and this intersection must contain just one element, the least upper bound of x and y. □

Note that one may apply a simple inductive argument (cf. e.g. [87] Chap. I, § 1, no. 1, or [362]) to transform the Riesz decomposition property into a similar decomposition property involving sums with more than two terms:

$$\begin{aligned} u=\sum_{i=1}^{n} u_i=\sum_{j=1}^{m} v_j, \quad \text{all } u_i, v_j \geqq 0, \\ \Rightarrow \exists w_{ij} \geqq 0:\ u_i=\sum_{j=1}^{m} w_{ij}, \quad v_j=\sum_{i=1}^{n} w_{ij}. \end{aligned} \tag{3.3}$$

Proposition II.3.3. *A compact convex set K is a simplex iff for any two proper convex combinations on K representing the same point, say*

$$x=\sum_{i=1}^{n} \lambda_i x_i=\sum_{j=1}^{m} \mu_j y_j, \tag{3.4}$$

there exists a new convex combination $x=\sum_{i,j} v_{ij}z_{ij}$ *on K such that*

$$x_i=\sum_{j=1}^{m}\lambda_i^{-1}v_{ij}z_{ij},\qquad y_j=\sum_{i=1}^{n}\mu_j^{-1}v_{ij}z_{ij}. \tag{3.5}$$

Proof. 1. If K is a simplex, then one may apply an argument similar to the proof of Proposition II.2.5 to translate the statement (3.3) about sums of elements from the (positive) cone $\tilde{K}$ into the corresponding statement about convex combinations of elements from the base K.

2. If K has the property stated in the proposition, then one may translate this property back into (3.3), which reduces to the original Riesz decomposition property for $n=m=2$. By Proposition II.3.1 and Proposition II.3.2, K is a simplex. □

Corollary II.3.4. *If x is a point of a simplex K, then the set* $M_x^+(K)'$ *of simple measures in* $M_x^+(K)$ *is directed in the ordering of Choquet.*

Proof. Application of Proposition II.3.3 and Corollary I.3.4. □

In the sequel we shall use the symbol $Z(K)$ to denote the linear space of all boundary measures on a compact convex set K and the symbol $Z_1^+(K)$ to denote the set of positive and normalized measures in $Z(K)$. For a given $x\in K$ we shall denote by $Z_x(K)$ the subspace of $Z(K)$ consisting of all measures μ which *represent* x, in that $\mu(a)=a(x)$ for all $a\in A(K)$. Also we shall use the symbol $Z_x^+(K)$ to denote the set of all positive (and necessarily normalized) measures in $Z_x(K)$. By definition $Z(K)$ is closed under the operation $\mu\rightsquigarrow|\mu|$; hence it is a *lattice subspace* of $M_{\mathbb{R}}(K)$. Note also that $Z_1^+(K)$ is a *face* of $M_1^+(K)$ and that $Z_x^+(K)$ is a *face* of $M_x^+(K)$ (These faces are not w^*-closed in general.)

We shall need the following result which is a simple consequence of the Choquet-Bishop-de Leeuw Theorem.

Lemma II.3.5. *Let K be a compact convex set. Then the restriction map* $\eta: C_{\mathbb{R}}(K)^*\to A(K)^*$ *maps the subspace* $Z(K)$ *of* $M_{\mathbb{R}}(K)=C_{\mathbb{R}}(K)^*$ *onto* $A(K)^*$, *and the mapping is* $1-1$ *on* $Z(K)$ *iff* $Z_x^+(K)$ *reduces to a single point for every* $x\in K$.

Proof. 1. Every element of the (base-norm) space $A(K)^*$ is of the form $q=\lambda_1 p_1-\lambda_2 p_2$, where p_1,p_2 are states on $A(K)$ (Prop. II.1.4).

Let p_1, p_2 the point states at $x_1, x_2 \in K$ (see § 2), and let μ_1, μ_2 be positive and normalized boundary measures with barycenter x_1, x_2, respectively. Then $\eta(\mu_i) = p_i$ for $i = 1, 2$. Writing $\mu = \lambda_1 \mu_1 - \lambda_2 \mu_2$, we obtain $\mu \in Z(K)$ and $\eta(\mu) = q$. The surjectivity is proved.

2. Assume next that every point x is the barycenter of a unique positive and normalized boundary measure. We shall prove that the zero-measure is the only measure $\mu \in Z(K)$ for which $\eta(\mu) = 0$. The equation $\eta(\mu) = 0$ means that $\mu(a) = 0$ for all $a \in A(K)$, or equivalently that $\mu^+(a) = \mu^-(a)$ for all $a \in A(K)$. In particular $\mu^+(1_K) = \mu^-(1_K)$, and without lack of generality we may assume that μ^+ and μ^- are normalized; hence $\mu^+, \mu^- \in Z_1^+(K)$. Now it follows that μ^+ and μ^- have common barycenter, and by assumption they are equal. Hence $\mu = 0$, and the injectivity is proved. □

Theorem II.3.6. *(Choquet) A compact convex set K is a simplex iff every point in K is the barycenter of a unique positive and normalized boundary measure.*

Proof. 1. Assume first that K is a simplex and consider an arbitrary $x \in K$. We claim that $M_x^+(K)$ is directed in Choquet's ordering, and from this the uniqueness will follow. To prove the directedness we consider two measures $\mu_1, \mu_2 \in M_x^+(K)$, and we define for every finite sequence $g_1, \ldots, g_n \in P(K)$ and every $\varepsilon > 0$ the w^*-closed set $A(\varepsilon, g_1, \ldots, g_n)$ consisting of all measures $\nu \in M_x^+(K)$ such that

$$\nu(g_i) \geqq \mu_j(g_i) = \varepsilon, \quad \text{for } i = 1, \ldots, n;\ j = 1, 2.$$

The sets $A(\varepsilon, g_1, \ldots, g_n)$ are non-void by the density of the set $M_x(K)'$ of simple measures in $M_x^+(K)'$ (Prop. I.2.3) and the fact that the set $M_x^+(K)'$ is directed in Choquet's ordering (Cor. II.3.4). The collection of sets $A(\varepsilon, g_1, \ldots, g_n)$ has the finite intersection property. By w^*-compactness of $M_x^+(K)$ these sets have a common member ν, which must be an upper bound of μ_1 and μ_2 with respect to Choquet's ordering.

2. Assume next that every point in K is the barcyenter of a unique positive and normalized boundary measure. By Lemma II.3.5, η is a $1-1$ map of $Z(K)$ onto $A(K)^*$, and it is seen to be an order isomorphism $(\eta(\mu) \geqq 0 \Leftrightarrow \mu \geqq 0)$. Generally $Z(K)$ is a vector lattice, and it follows that $A(K)^*$ is also a vector lattice in the present situation. By definition K is a simplex, and the proof is complete. □

For every element x of a simplex K we shall denote the unique element of $Z_x^+(K)$ by μ_x, and we observe that it follows from Proposition I.3.1 and Corollary I.3.6 that

$$\mu_x(f)=\hat{f}(x), \quad \text{all } f\in P(K). \tag{3.6}$$

Theorem II.3.7. *(Choquet-Meyer) If K is a compact convex set, then the following statements are equivalent:*

(i) K is a simplex.

(ii) $\mu(f)=\hat{f}(x)$ for all $x\in K$, $\mu\in Z_x^+(K)$, $f\in P(K)$,

(iii) $\hat{f}+\hat{g}=\widehat{f+g}$ for all $f,g\in P(K)$,

(iv) $\hat{f}$ is an affine function for all $f\in P(K)$.

The *proof* proceeds in two cycles (i) ⇒ (ii) ⇒ (iii) ⇒ (i), (i) ⇒ (iv) ⇒ (ii).

(i) ⇒ (ii) is trivial by (3.6).

(ii) ⇒ (iii). If $f,g\in P(K)$, then $f+g\in P(K)$; hence for arbitrary $x\in K$ and $\mu\in Z_x^+(K)$:

$$\hat{f}(x)+\hat{g}(x)=\mu(f)+\mu(g)=\mu(f+g)=\widehat{f+g}(x).$$

(iii) ⇒ (i). Let $x\in K$ and consider the functional $f\rightsquigarrow\hat{f}(x)$ on $P(K)$. It is positively homogeneous (Prop. I.1.6), and by assumption it is also additive. Hence $\varphi(f-g)=\hat{f}(x)-\hat{g}(x)$ defines a linear functional on the dense linear subspace $P(K)-P(K)$ of $C_{\mathbb{R}}(K)$. By elementary properties of envelopes (Prop. I.1.6) we shall have

$$|\varphi(f-g)|\leqq\|f-g\|.$$

Hence φ has a unique extension to a linear functional $\mu\in C_{\mathbb{R}}(K)^* =M_{\mathbb{R}}(K)$ of norm 1. Since $\varphi(1)=1$ we shall have $\mu\in M_1^+(K)$, and it follows by the definition of μ that $\mu(f)=\hat{f}(x)$ for all $f\in P(K)$. By Proposition I.3.1 and Corollary I.3.6 this implies $\nu\prec\mu$ for every $\nu\in M_x^+(K)$. Hence μ is the unique maximal element of $M_x^+(K)$, and by Theorem II.3.6 K is a simplex.

(i) ⇒ (iv). Clearly $x\rightsquigarrow\mu_x$ is an affine mapping, and $\hat{f}(x)=\mu_x(f)$ by (3.6). Hence $\hat{f}$ is affine.

(iv) ⇒ (ii). If $f\in P(K)$, then $\hat{f}$ is u.s.c. and affine. By Corollary I.1.4 there is a net $\{a_\alpha\}$ from $A(K)$ such that $a_\alpha\searrow\hat{f}$, and so we shall have for an arbitrary $x\in K$ and an arbitrary $\mu\in Z_x^+(K)$

$$\mu(f)=\mu(\hat{f})=\lim_\alpha\mu(a_\alpha)=\lim_\alpha a_\alpha(x)=\hat{f}(x).$$

This completes the proof. □

Remark. It is seen by inspection of the above proof that one may replace the cone $P(K)$ in Theorem II.3.7 by any dense subcone, e.g. the cone of all functions $a_1 \vee \cdots \vee a_n$ where $a_i \in A(K)$, or even $a_i \in A(K;E)$, $i=1,\dots,n$.

The equivalence (i)$\Leftrightarrow$(iv) of Theorem II.3.7 (or rather its dual version involving $\check{f}$ and $-P(K)$) is sometimes referred to as the „stable table principle". It generalizes the old fact that only a three-legged table will be stable.

In Theorem II.3.7 we characterized simplexes by properties involving continuous convex and affine functions. In our next theorem we shall characterize simplexes by properties involving upper semi-continuous convex and affine functions.

Theorem II.3.8. *If K is a compact convex set, then the following statements are equivalent:*

(i) *K is a simplex.*

(ii) *$\hat{f}$ is affine for every u.s.c. convex function $f: K \to [-\infty, \infty[$.*

(iii) *Any two members of the cone of u.s.c. affine functions from K into $[-\infty, \infty[$ has a l.u.b. in this cone.*

Proof. (i)$\Rightarrow$(ii). Assume (i) and consider an u.s.c. convex function $f: K \to [-\infty, \infty[$. By Proposition I.5.1 there is a net $\{f_\alpha\}$ from $P(K)$ such that $f_\alpha \to f$. Now we can proceed as in the second part of the proof of Proposition I.3.1 to prove that $\hat{f}_\alpha \searrow \hat{f}$. By Theorem II.3.7 $\hat{f}_\alpha$ is affine for every α, and hence $\hat{f}$ is affine as well.

(ii)$\Rightarrow$(iii). Assume (ii) and consider two u.s.c. affine functions $f, g: K \to [-\infty, \infty[$. Clearly $f \vee g$ is u.s.c. and convex, and so $\widehat{f \vee g}$ is u.s.c. and affine.

(iii)$\Rightarrow$(i). Assuming (iii) we conclude that every finite sequence $f_1, \dots, f_n$ from the cone of u.s.c. affine functions from K into $[-\infty, \infty[$ has a l.u.b. in this cone, and by the definition of upper envelopes it is equal to $\widehat{f_1 \vee \cdots \vee f_n}$. In particular we may take $a_1, \dots, a_n \in A(K)$, and by the Remark to Theorem II.3.7, K is a simplex. The proof is complete. □

Corollary II.3.9. *Let K be a simplex and let $f: K \to [-\infty, \infty[$ be an u.s.c. convex function and $g: K \to]-\infty, \infty]$ a l.s.c. concave function. If $f \leqq g$ then $\hat{f} \leqq \check{g}$, and if $f < g$ then $\hat{f} < \check{g}$.*

Proof. By Theorem II.3.8 the function $\check{g} - \hat{f}$ is l.s.c. and affine. Hence it assumes its infimum value at some point $x \in \partial_e K$ (by the Krein-Milman Theorem).

By Proposition I.4.1 we obtain

$$\inf_{y\in K}(\check{g}(y)-\hat{f}(y))=\check{g}(x)-\hat{f}(x)=g(x)-f(x)$$

and the desired results follow. □

If K is a simplex and $a_i, b_j \in A(K)$ with $a_i < b_j$ for $i,j=1,2$, then $\widehat{a_1 \vee a_2} < \widecheck{b_1 \wedge b_2}$ by virtue of Corollary II.3.9. Now it follows by a standard argument based on Hahn-Banach separation in product space (cf. the proof of Proposition I.1.2) that there exists an $h \in A(K)$ such that

$$a_1 \vee a_2 < h < b_1 \wedge b_2. \tag{3.7}$$

The property stated above, is usually referred to as the *weak Riesz interpolation property*.

We shall see that the space $A(K)$ over a simplex K also enjoys the usual (or "strong") Riesz interpolation property. However, this is less trivial, and it depends on the following:

Theorem II.3.10. *(Edwards) Let K be a simplex and let $f: K \to [-\infty, \infty[$ be an u.s.c. convex function and $g: K \to]-\infty, \infty]$ a l.s.c. concave function. If $f \leqq g$, then there exists an $h \in A(K)$ such that $f \leqq h \leqq g$.*

Proof. By an inductive argument we shall construct a sequence $\{h_n\}_{n=0}^{\infty}$ from $A(K)$ such that for $n \geqq 1$

$$f-1 < h_0 < g+1, \tag{3.8}$$

$$(h_{n-1}-2^{-n}) \vee (f-2^{-n}) < h_n < (h_{n-1}+2^{-n}) \wedge (g+2^{-n}), \tag{3.9}$$

and we observe that this will establish the theorem since $\{h_n\}$ will be a Cauchy sequence converging to a function $h \in A(K)$ for which $f \leqq h \leqq g$.

For $n=0$ we have $\widehat{f-1} < \widecheck{g+1}$ by virtue of Corollary II.3.9. By Hahn-Banach separation in product space there is an $h_0 \in A(K)$ satisfying (3.8).

In the n-th step we use the induction-hypothesis to verify that

$$(h_{n-1}-2^{-n}) \vee (f-2^{-n}) < (h_{n-1}+2^{-n}) \wedge (g+2^{-n}).$$

By Corollary II.3.9 and by Hahn-Banach separation in product space there is an $h_n \in A(K)$ satisfying (3.9), and the induction is complete. □

Corollary II.3.11. *If K is a compact convex set, then the following statements are equivalent:*

(i) *K is a simplex.*

(ii) *$A(K)$ has the Riesz interpolation property.*

(iii) *$A(K)$ has the weak Riesz interpolation property.*

Proof. (i) $\Rightarrow$ (ii). If $a_i, b_j \in A(K)$ and $a_i \leqq b_j$ for $i,j=1,2$, then there exists by Theorem II.3.10 a function $h \in A(K)$ such that $a_1 \vee a_2 \leqq h \leqq b_1 \wedge b_2$.

(ii) $\Rightarrow$ (iii). If $a_i, b_j \in A(K)$ and $a_i < b_j$ for $i,j=1,2$, then $a_1 \vee a_2 + \varepsilon \leqq b_1 \wedge b_2 - \varepsilon$ for some $\varepsilon > 0$ by continuity and compactness. By Riesz interpolation property there is an $h \in A(K)$ such that

$$a_1 \vee a_2 < (a_1 + \varepsilon) \vee (a_2 + \varepsilon) \leqq h \leqq (b_1 - \varepsilon) \wedge (b_1 - \varepsilon) < b_1 \wedge b_2.$$

(iii) $\Rightarrow$ (i). By finite induction we can prove that if $a_i, b_j \in A(K)$ and $a_i < b_j$ for $i=1,\dots,n$, and $j=1,2$, then there exists an $h \in A(K)$ such that

$$a_1 \vee \cdots \vee a_n < h < b_1 \wedge b_2.$$

Hence the set of all $b \in A(K)$ for which $b > a_1 \vee \cdots \vee a_n$, is directed downwards. It follows that $a_1 \vee \cdots \vee a_n$ is affine; and by the Remark to Theorem II.3.7, K will be a simplex. $\square$

Theorem II.3.12. *Let K be a compact convex set and consider the following three statements:*

(i) *K is a simplex.*

(ii) *For every compact subset C of $\partial_e K$ and every $f \in C_{\mathbb{R}}(C)$ there exists an $a \in A(K)$ such that $a|c = f$ and $\|a\| = \|f\|$.*

(iii) *There exists a $\lambda \in \mathbb{R}^+$ such that for every compact subset C of $\partial_e K$ the set of all $a|c$ where $a \in A(K)$, $\|a\| \leqq \lambda$, is dense in the unit ball of $C_{\mathbb{R}}(C)$.*

Generally (i) *implies* (ii), *and* (ii) *implies* (iii). *If K is metrizable, then all three statements are equivalent.*

Proof. 1. We shall first prove that (i) implies (ii), and to this end we consider an arbitrary compact subset C of $\partial_e K$ and an arbitrary function f in $C_{\mathbb{R}}(C)$. Let $\alpha = \sup_{x \in C} f(x)$, $\beta = \inf_{x \in C} f(x)$, and define two functions g, k such that $g(x) = k(x) = f(x)$ for $x \in C$, while $g(x) = \alpha$ for $x \in K \backslash C$ and $k(x) = \beta$ for $x \in K \backslash C$.

It is easily verified that k is u.s.c. and convex while g is l.s.c. and concave. By Theorem II.3.10 there exists an $a \in A(K)$ such that $k \leqq a \leqq g$.

Now $a|c=f$ and $\|a\|=\alpha\vee(-\beta)=\|f\|$. This completes the first part of the proof.

2. Trivially (ii) implies (iii).

3. Assume finally that K is a metrizable compact convex set with the property (iii). To prove that K is a simplex, we consider two measures $\mu, \nu \in Z_x(K)$ for some $x\in K$ and we shall show that $\mu=\nu$.

Let $g\in C_{\mathbb{R}}(C)$, $\|g\|\leqq 1$, and let $\varepsilon>0$ be arbitrary. Since K is metrizable, $\partial_e K$ is a G_δ-set and $\mu(\partial_e K)=\nu(\partial_e K)=1$. By regularity there is a compact subset C of $\partial_e K$ such that

$$\mu(C)>1-\varepsilon, \qquad \nu(C)>1-\varepsilon. \tag{3.10}$$

By assumption there exists an $a\in A(K)$ such that $\|a\|\leqq\lambda$ and

$$|a(x)-g(x)|<\varepsilon, \quad \text{for all } x\in C. \tag{3.11}$$

Using (3.10), (3.11), and the equality $\mu(a)=\nu(a)$, we obtain

$$|\mu(g)-\nu(g)|\leqq|\mu(g)-\mu(a)|+|\nu(a)-\nu(g)|\leqq(4+2\lambda)\varepsilon.$$

Since $\varepsilon>0$ is arbitrary, we shall have $\mu(g)=\nu(g)$, and the proof is complete. □

It follows from Theorem II.3.12 that if K is a simplex and $\partial_e K$ is closed, then every $f\in C_{\mathbb{R}}(\partial_e K)$ can be extended with preservation of norm to a function in $A(K)$. However, this particular result can be obtained more directly without the use of Edwards' Theorem:

Proposition II.3.13. *If K is a simplex and $\partial_e K$ is closed, then every $f\in C_{\mathbb{R}}(\partial_e K)$ can be uniquely extended to a function $\tilde{f}\in A(K)$; in fact $\tilde{f}(x)=\mu_x(f)$ for all $x\in K$ and $\|\tilde{f}\|=\|f\|$.*

Proof. Since $\partial_e K$ is closed, a measure $\mu\in M_1^+(K)$ will be a boundary measure iff $\operatorname{Supp}(\mu)\subset\partial_e K$. (Prop. I.4.6 and formula (4.10).) It follows that $\mu \rightsquigarrow x_\mu$ is a $1-1$ bicontinuous mapping of the vaguely compact set $M_1^+(\partial_e K)$ onto K. The inverse map is just $x\rightsquigarrow\mu_x$, and so the function $\tilde{f}: x\rightsquigarrow\mu_x(f)$ is continuous. Clearly $\tilde{f}$ is affine and $\|\tilde{f}\|\leqq\|f\|$. The uniqueness is trivial, and the proof is complete. □

The extreme boundary of a simplex need not be closed, and then the conclusions of Proposition II.3.13 are inexact. (We shall see examples to this effect in a subsequent section.) However, the following weaker proposition will subsist in the general case:

Proposition II.3.14. *Let K be a simplex and f a uniformly continuous real valued function on $\partial_e K$. Then f can be extended to an affine Borel function $\tilde{f}$ for which the barycenter formula is valid. In fact we may choose $\tilde{f}(x)=\mu_x(\bar{f})$ where $\bar{f}$ is the continuous extension of f to $\overline{\partial_e K}$; then $\|\tilde{f}\|=\|f\|$, and $\tilde{f}$ is uniform limit of a sequence $\{h_{1,n}-h_{2,n}\}_{n=1}^{\infty}$, where $h_{i,n}$ are u.s.c. affine functions for $i=1,2$; $n=1,2,\ldots$. If K is metrizable, then this is the only possible choice of $\tilde{f}$.*

Proof. We extend $\bar{f}$ to a continuous function defined on K by Tietze's Theorem, and then we apply Proposition I.1.1 to construct a sequence $\{g_{1,n}-g_{2,n}\}_{n=1}^{\infty}$ with $g_{1,n}, g_{2,n}\in P(K)$ such that

$$|\bar{f}(y)-(g_{1,n}(y)-g_{2,n}(y))|\leqq \frac{1}{n} \tag{3.12}$$

for all $y\in\overline{\partial_e K}$ and for $n=1,2,\ldots$.

By Theorem II.3.7 the functions $h_{i,n}=\hat{g}_{i,n}$ are u.s.c. and affine for $i=1,2$; $n=1,2,\ldots$. Making use of (3.6), we obtain

$$\mu_x(g_{1,n}-g_{2,n})=h_{1,n}(x)-h_{2,n}(x) \tag{3.13}$$

for all $x\in K$ and for $n=1,2,\ldots$.

Since $\operatorname{Supp}(\mu_x)\subset\overline{\partial_e K}$ for every $x\in K$ (Prop. I.4.6), the function $\tilde{f}$ is well defined by $\tilde{f}(x)=\mu_x(\bar{f})$, and it follows from (3.12) and (3.13) that

$$\|\tilde{f}-(h_{1,n}-h_{2,n})\|\leqq\frac{1}{n}\quad\text{for } n=1,2,\ldots.$$

This proves that $\tilde{f}$ is an affine Borel function of the type stated in the proposition. In particular, the barycenter formula is valid for $\tilde{f}$ since it is valid for every u.s.c. affine function. (Cf. Cor. I.1.4.) Clearly $\|\tilde{f}\|\leqq\|f\|$, and $\tilde{f}$ is an extension of f since $\mu_x=\varepsilon_x$ for $x\in\partial_e K$.

Assume finally that K is metrizable and that f' is some Borel function which extends f and for which the barycenter formula is valid. Then $\partial_e K$ is a G_δ-set and for every $x\in K$

$$f'(x)=\int_K f'\,d\mu_x=\int_{\partial_e K} f\,d\mu_x.$$

Now the uniqueness follows, and the proof is complete. □

Remark. If we replace the word "Borel" with "Baire" in the above proposition, then we can prove uniqueness even in the nonmetrizable case (cf. Th. I.4.14), but the existence proof will break down.

Corollary II.3.15. *If K is a simplex, then the map $x \rightsquigarrow \mu_x$ of K into $M_1^+(K)$ is vaguely Borel, i.e. $x \rightsquigarrow \mu_x(f)$ is Borel measurable for every $f \in C_{\mathbb{R}}(K)$.*
theorem:

Proof. By Proposition II.3.14 the function $x \rightsquigarrow \mu_x(f)$ is uniform limit of a sequence of differences of semi-continuous functions. Hence it is Borel measurable. □

We close this section by the following version of the uniqueness theorem.

Proposition II.3.16. *A compact convex set K is a simplex iff every point in K is the barycenter of a unique simplicial boundary measure.*

The *proof* is a straightforward application of Theorem II.3.6 and Theorem I.6.14. □

* We shall see that the "non-metrizable" pathologies of boundary measures can occur even for simplexes. In fact we shall show that the compact convex set constructed in Proposition I.4.15, is a simplex.

Proposition II.3.17. *There exists a simplex K with a point z such that*
(i) *$\partial_e K$ is a Borel subset of K and $\partial_e K$ is a discrete space in relative topology.*
(ii) $\mu_z(\partial_e K)=0$.

Proof. We consider the set K constructed in the proof of Proposition I.4.15, and we shall make free use of the notation of that proof.

Observe first that $M_1^+(Y)$ is regularly embedded in $M_{\mathbb{R}}(Y)$, that $K=\varphi(M_1^+(Y))$ is regularly embedded in $E=M_{\mathbb{R}}(Y)/N$, and that there is a bijection $f \rightsquigarrow a_f$ of the set

$$N_\perp=\{f \in C_{\mathbb{R}}(Y) \mid \nu(f)=0 \text{ all } \nu \in N\} \tag{3.14}$$

onto $A(K)$, satisfying

$$a_f(\varphi(\mu))=\mu(f), \quad \text{all } \mu \in M_1^+(Y). \tag{3.15}$$

(This invokes an elementary result on the dual of a quotient space, cf. e.g. [86]; cf. also formula (4.27).)

We shall prove that K is a simplex by showing that it enjoys the "stable table" property (Theorem II.3.7, property (iv)). By the Remark to Theorem II.3.7 it suffices to consider a continuous convex function on K of the special form $F=a_{f_1}\vee\cdots\vee a_{f_n}$, where $f_1,\ldots,f_n\in N_\perp$.

We define a function $g: Y\to\mathbb{R}$ by first writing

$$g(r_\alpha)=f_1(r_\alpha)\vee\cdots\vee f_n(r_\alpha),\qquad g(t_\alpha)=f_1(t_\alpha)\vee\cdots\vee f_n(t_\alpha),\tag{3.16}$$

and then

$$g(s_\alpha)=\tfrac{1}{2}(g(r_\alpha)+g(t_\alpha)).\tag{3.17}$$

We claim that g is u.s.c. For the proof of this claim it suffices to consider points of the type s_α, since the other points of Y are isolated. By the continuity of $f_1,\ldots,f_n$ and by the definition of the ("porcupine") topology on Y there exists for every $\alpha\in[0,1]$ and every $\varepsilon>0$ a number $\delta>0$ such that

$$|f_i(y)-f_i(s_\alpha)|<\varepsilon,\quad \text{for } i=1,\ldots,n,\tag{3.18}$$

whenever $y\in Y_\beta$ and $0<|\alpha-\beta|<\varepsilon$.

By (3.16) and (3.18) we obtain for all β such that $0<|\alpha-\beta|<\varepsilon$:

$$g(r_\beta)<f_1(s_\alpha)\vee\cdots\vee f_n(s_\alpha)+\varepsilon,$$

and similarly

$$g(t_\beta)<f_1(s_\alpha)\vee\cdots\vee f_n(s_\alpha)+\varepsilon.$$

Hence for all β such that $0<|\alpha-\beta|<\varepsilon$, we shall have

$$\begin{aligned} g(s_\beta) &< \max_{1\leqq i\leqq n}\,[\tfrac{1}{2}\,(f_i(r_\alpha)+f_i(t_\alpha)]+\varepsilon\\ &\leqq \tfrac{1}{2}\,[\max_{1\leqq i\leqq n} f_i(r_\alpha)+\max_{1\leqq i\leqq n} f_i\,(t_\alpha)]+\varepsilon\\ &= g(s_\alpha)+\varepsilon\,. \end{aligned}$$

This proves that g ist u.s.c.

Clearly $v(g)=0$ for all $v\in N$. Hence there is a function $G: K\to\mathbb{R}$ satisfying

$$G(\varphi(\mu))=\mu(g)\,.\tag{3.19}$$

It follows from the upper semi-continuity of g that the function $G_0:\mu\rightsquigarrow\mu(g)$ is upper semi-continuous on $M_1^+(Y)$. Now $G_0=G\circ\varphi$, and so it follows by elementary properties of the quotient topology that G is u.s.c.

By the definition (3.16) we shall have

$$\begin{aligned} G(\varphi(\varepsilon_{r_\alpha}))&=g(r_\alpha)=F(\varphi(\varepsilon_{r_\alpha}))\,,\\ G(\varphi(\varepsilon_{t_\alpha}))&=g(t_\alpha)=F(\varphi(\varepsilon_{t_\alpha}))\,. \end{aligned}$$

Hence G is an u.s.c. affine function on K which coincides with F on $\partial_e K$. It follows that $G=\hat{F}$. By Theorem II.3.7 this completes the proof that K is a simplex.

The property (i) is already verified in the proof of Proposition I.4.15.

Finally we define $\tilde{\lambda}$ to be the direct image in Y of Lebesgue measure λ on $[0,1]$ under the map $s:\alpha \rightsquigarrow s_\alpha$, and we define $z=\varphi(\tilde{\lambda})$. It follows by an easy modification of the last verification in the proof of Proposition I.4.15 that for every $\varepsilon>0$, the norm of the purely atomic part of μ_z is less than ε. (In the previous proof we had chosen $\varepsilon=\frac{1}{2}$.) Now μ_z is non-atomic, and since $\partial_e K$ is a discrete space, the regularity of μ_z will imply $\mu_z(\partial_e K)=0$. (Cf. the proof of Prop. I.4.16.) □

It is of interest to note that the requirement to metrizability can not be avoided in Theorem II.3.12. To prove this we shall need the following:

Lemma II.3.18. *Let K be the simplex constructed in Proposition* II.3.17 *together with the point z. Then*

(i) *There exist functions $a_1, a_2 \in A(K)$ and points $x_1, x_2 \in \partial_e K$ such that $a_i(z)=a_i(x_1)=a_i(x_2)=0$ for $i=1,2$, and $\mu_z(a_1 \vee a_2) \geqq 1$.*

(ii) *For every finite subset C of $\partial_e K$ and every real valued function h on C and every real number γ such that $|\gamma| \leqq \|h\|$, there exists an $a \in A(K)$ for which $a|c=h$, $\|a\|=\|h\|$ and $a(z)=\gamma$.*

Proof. 1. Let $\alpha \in [0,1]$ be arbitrary and define $x_1=\varphi(\varepsilon_{r_\alpha})$, $x_2=\varphi(\varepsilon_{t_\alpha})$. Let $g_1 \in C_{\mathbb{R}}([0,1])$ be chosen such that $\lambda(g_1)=g_1(\alpha)=0$ and $\lambda(|g_1|) \geqq 1$. Writing $g_2=-g_1$ we obtain $\lambda(g_2)=g_2(\alpha)=0$ and $\lambda(g_1 \vee g_2)=\lambda(|g_1|) \geqq 1$.

Next we define $\tilde{g}_i : Y \to \mathbb{R}$ by writing

$$\tilde{g}_i(r_\beta)=\tilde{g}_i(s_\beta)=\tilde{g}_i(t_\beta)=g_i(\beta), \quad \text{for } \beta \in [0,1]; i=1,2. \tag{3.20}$$

Clearly $\tilde{g}_1, \tilde{g}_2 \in N_\perp$ and so there are functions $a_1, a_2 \in A(K)$ such that

$$a_i(\varphi(\mu))=\mu(\tilde{g}_i), \quad \text{for all } \mu \in M_1^+(Y); i=1,2. \tag{3.21}$$

(Cf. formula (3.15).)

In particular we obtain for $i=1,2$:

$$a_i(x_1)=\varepsilon_{r_\alpha}(\tilde{g}_i)=0, \qquad a_i(x_2)=\varepsilon_{t_\alpha}(\tilde{g}_i)=0, \tag{3.22}$$

and

$$\mu_z(a_i)=a_i(z)=a_i(\varphi(\tilde{\lambda}))=\tilde{\lambda}(\tilde{g}_i)=\lambda(g_i)=0. \tag{3.23}$$

Now let $\tilde{\Lambda}$ be the direct image of the measure $\tilde{\lambda}$ under the map $\psi : y \rightsquigarrow \varepsilon_y$ of Y into $M_1^+(Y)$. Hence $\tilde{\Lambda}$ is measure on $M_1^+(Y)$, and its direct image $\varphi\tilde{\Lambda}$ is a positive and normalized measure on K.

We *claim* that $\varphi\tilde{\Lambda}$ represents the point $z=\varphi(\tilde{\lambda})$.

To prove this in full detail, we choose an arbitrary $b \in N_{\perp}$, and we denote by a_b the corresponding function in $A(K)$. Thus

$$a_b(\varphi(\mu)) = \mu(b) = b^*(\mu), \quad \text{all } \mu \in M_1^+(Y). \tag{3.24}$$

Now

$$\begin{aligned}[\varphi\tilde{\Lambda}](a_b) &= \int (a_b \circ \varphi)\, d\tilde{\Lambda} = \int b^*\, d(\psi\tilde{\lambda}) \\ &= \int (b^* \circ \psi)\, d\tilde{\lambda} = \tilde{\lambda}(b) = a_b(\varphi(\tilde{\lambda})),\end{aligned} \tag{3.25}$$

and the claim is proved.

We know that $M_z^+(K)$ is *directed* in Choquet's ordering (cf. the proof of Th. II.3.6). Hence $\varphi\tilde{\Lambda} \prec \mu_z$. By an argument similar to (3.25) applied to the function $a_1 \vee a_2 \in P(K)$, we obtain

$$\begin{aligned}\mu_z(a_1 \vee a_2) \geqq [\varphi\tilde{\Lambda}](a_1 \vee a_2) &= \int (a_1 \circ \varphi) \vee (a_2 \circ \varphi)\, d\tilde{\Lambda} \\ &= \tilde{\lambda}(\tilde{g}_1 \vee \tilde{g}_2) = \lambda(g_1 \vee g_2) \geqq 1 .\end{aligned}$$

Together with (3.22) and (3.23) this completes the first part of the proof.

2. To prove (ii) we consider a finite subset C of $\partial_e K$, a real valued function h on C and a number γ such that $|\gamma| \leqq \|h\|$. To simplify the notation we assume that C can be expressed in the following way for some finite subset C_0 of $[0,1]$:

$$C = \{\varepsilon_{r_\alpha} \mid \alpha \in C_0\} \cup \{\varepsilon_{s_\alpha} \mid \alpha \in C_0\} . \tag{3.26}$$

(This can always be achieved by some trivial norm preserving extension of h.)

Now we define a function k_0 on C_0 by writing

$$k_0(\alpha) = \tfrac{1}{2}\left[h(\varphi(\varepsilon_{r_\alpha}) + h(\varphi(\varepsilon_{t_\alpha}))\right], \quad \text{for } \alpha \in C_0 . \tag{3.27}$$

By an elementary construction we can extend k_0 to a function $k \in C_{\mathbb{R}}([0,1])$ such that $\|k\| = \|k_0\| = \|h\|$ and $\lambda(k) = \gamma$. (We can take a suitable piecewise linear function.)

Next we define $\tilde{k}: Y \to \mathbb{R}$ by writing

$$\tilde{k}(r_\alpha) = h(\varphi(\varepsilon_{r_\alpha})), \qquad \tilde{k}(s_\alpha) = k_0(\alpha), \qquad \tilde{k}(t_\alpha) = h(\varphi(\varepsilon_{t_\alpha})) \tag{3.28}$$

for all $\alpha \in C_0$, and

$$\tilde{k}(r_\alpha) = \tilde{k}(s_\alpha) = \tilde{k}(t_\alpha) = k(\alpha) \tag{3.29}$$

for all $\alpha \in [0,1] \setminus C_0$.

Clearly $k \in N_{\perp}$, and so there is an $a \in A(K)$ such that

$$a(\varphi(\mu)) = \mu(\tilde{k}), \qquad \text{for all } \mu \in M_1^+(Y) . \tag{3.30}$$

By (3.28) and (3.29) we obtain

$$a(\varphi(\varepsilon_{r_\alpha})) = h(\varphi(\varepsilon_{r_\alpha})), \qquad a(\varphi(\varepsilon_{t_\alpha})) = h(\varphi(\varepsilon_{t_\alpha})) ,$$

for all $\alpha \in C_0$.

Hence $a(y)=h(y)$ for all $y\in C$, or otherwise stated $a|c=h$.
Clearly $\|a\|=\|\tilde{k}\|=\|h\|$, and finally by (3.30)

$$a(z)=a(\varphi(\tilde{\lambda}))=\tilde{\lambda}(\tilde{k})=\lambda(k)=\gamma\,.$$

The proof is complete. □

Proposition II.3.19. *There exists a compact convex set K' which is not a simplex, but on which every continuous real valued function f defined on a compact subset C of $\partial_e K'$ can be extended with preservation of norm to a function in $A(K)$.*

Proof. Let K be the simplex of Proposition II.3.17, and consider two points $x_1, x_2\in\partial_e K$ together with two functions $a_1, a_2\in A(K)$ such that requirement (i) of Lemma II.3.18 is satisfied.

We shall deform K into a non-simplex by a construction which identifies z with the mid-point of the segment $[x_1, x_2]$. This construction is essentially the same as the one used in the proof of Proposition I.4.15, but we are now in the position to make certain shortcuts, since we can apply the general embedding technique of § 2 of this chapter.

Let B be the space of all $a\in A(K)$ satisfying

$$a(z)=\tfrac{1}{2}a(x_1)+\tfrac{1}{2}a(x_2)\,; \tag{3.31}$$

let K' be the state space of the order-unit space $(B, 1_K)$, and let $\psi': K\to K'$ be defined by $[\psi' x](a)=a(x)$ for all $a\in B$. We claim that ψ' is a bijection of $\partial_e K$ onto $\partial_e K'$.

To prove this claim, we first note that we have the following implication in which $y_1, y_2\in K$:

$$\psi'(y_1)=\psi'(y_2) \;\Rightarrow\; \exists\alpha\in\mathbb{R}: y_1-y_2=\alpha(z-\tfrac{1}{2}x_1-\tfrac{1}{2}x_2)\,. \tag{3.32}$$

In fact we may factorize $\psi'=\eta\circ\psi$, where ψ is the canonical isomorphism of K onto the state space of $(A(K), 1_K)$ and $\eta: A(K)^*\to B^*$ is the restriction map, and then obtain (3.32) since the kernel of η is the one-dimensional space spanned by the linear functional $\psi(z)-\frac{1}{2}\psi(x_1)-\frac{1}{2}\psi(x_2)$ on $A(K)$.

Let $y\in\partial_e K$ and assume that $\psi'(y)$ is non-extreme, say

$$\psi'(y)=\tfrac{1}{2}\psi'(u)+\tfrac{1}{2}\psi'(v)\,, \tag{3.33}$$

where $u, v\in K$, and $u\neq v$.

By (3.32)

$$y+\frac{\alpha}{2}x_1+\frac{\alpha}{2}x_2=\alpha z+\frac{1}{2}u+\frac{1}{2}v\,. \tag{3.34}$$

We assume first that $\alpha>0$, and divide through by $\alpha+1$. This gives an equality of two convex combinations. Since K is a simplex, the representing boundary measure is unique, and we obtain

$$\varepsilon_y+\frac{\alpha}{2}\varepsilon_{x_1}+\frac{\alpha}{2}\varepsilon_{x_1}=\alpha\mu_z+\frac{1}{2}\mu_u+\frac{1}{2}\mu_v\,.$$

The measures on the left hand side of this equation all vanish on $\partial_e K$, and it follows by property (ii) of Proposition II.3.14 that $\alpha=0$. Now formula (3.34) gives $y=\frac{1}{2}u+\frac{1}{2}v$, which is a contradiction since $y\in\partial_e K$. If the number α is negative, then we move the terms involving α to the opposite side in the equality (3.34) and proceed as before. Hence we have shown that ψ' maps $\partial_e K$ into $\partial_e K'$.

To prove that ψ' is a $1-1$ map of $\partial_e K$ onto $\partial_e K'$, it suffices to prove that for every $w\in\partial_e K'$ the set $F=\{y\in K\,|\,\psi'(y)=w\}$ reduces to a single point which is in $\partial_e K$. Now F is seen to be a closed face of K. Hence it is the closed convex hull of its extreme points, and these are in $\partial_e K$. Thus, either F reduces to a single point in $\partial_e K$, or F must contain at least two distinct points $y_1, y_2\in\partial_e K$. In the second case (3.32) gives

$$y_1+\frac{\alpha}{2}x_1+\frac{\alpha}{2}x_2=\alpha z+y_2$$

for some $\alpha\in\mathbb{R}$, and now we may proceed as before obtaining a contradiction unless $\alpha=0$; but this in turn contradicts the assumption $y_1\neq y_2$. Hence we have proved that ψ' is a bijection of $\partial_e K$ onto $\partial_e K'$.

Next we shall prove that K' is not a simplex.

Observe first that the measure

$$\mu=\tfrac{1}{2}\varepsilon_{\psi'(x_1)}+\tfrac{1}{2}\varepsilon_{\psi'(x_2)} \tag{3.35}$$

is a boundary measure which represents the point $\psi'(z)$. In fact, it follows from Theorem II.2.1 that there is a bijection $a \rightsquigarrow \tilde{a}$ of B onto $A(K')$ such that

$$\tilde{a}(\psi'(x))=a(x)\quad \text{all}\ \ x\in K\,, \tag{3.36}$$

and it is easily seen that $\mu(\tilde{a})=\tilde{a}(z)$ for all $a\in B$.

In a similar way one may prove that the direct image $\psi'\mu_z$ of μ_z by ψ' will represent the point $\psi'(z)$.

Clearly the functions a_1, a_2 belong to B, and we shall evaluate the measures μ and $\psi'\mu_z$ at the function $\tilde{a}_1\vee\tilde{a}_2\in P(K')$.

We obtain

$$\mu(\tilde{a}_1\vee\tilde{a}_2)=\tfrac{1}{2}[\tilde{a}_1\vee\tilde{a}_2](\psi'(x_1))+\tfrac{1}{2}[\tilde{a}_1\vee\tilde{a}_2](\psi'(x_2))=0 \tag{3.37}$$

$$\begin{aligned}[\psi'\mu_z](\tilde{a}_1\vee\tilde{a}_2)&=\int\tilde{a}_1\vee\tilde{a}_2\,d(\psi'\mu_z)\\&=\int(\tilde{a}_1\circ\psi')\vee(\tilde{a}_2\circ\psi')\,d\mu_z\geqq 1\,.\end{aligned} \tag{3.38}$$

Now if K' had been a simplex, then we should have $\psi'\mu_z \prec \mu$ by the maximality of μ and the directedness of $M^+_{\psi'(z)}(K')$ (cf. the proof of Th. II.3.5). But this is impossible by virtue of (3.37) and (3.38).

Finally we consider an arbitrary compact subset C of $\partial_e K'$ and a function $f \in C_{\mathbb{R}}(C)$. The set $C_0 = \{x \in K \mid \psi'(x) \in C\}$ is a closed, hence compact, subset of K. Also $C_0 \subset \partial_e K$, and it follows by the discreteness of $\partial_e K$ that C_0 is finite. Hence C is also finite. To simplify the writing we shall assume $\psi'(x_1), \psi'(x_2) \in C$. (This can always be achieved by a trivial extension of f.)

By property (iii) of Lemma II.3.17 there exists an $a \in A(K)$ such that $a|C_0 = f \circ \psi'$, $\|a\| = \|f\|$ and

$$a(z) = \tfrac{1}{2} f(\psi'(x_1)) + \tfrac{1}{2} f(\psi'(x_2)) \,. \tag{3.39}$$

Now $a \in B$, and the corresponding function $\tilde{a} \in A(K')$ satisfies

$$\tilde{a}(\psi'(y)) = a(y) = f(\psi'(y)) \,, \quad \text{for all } y \in C_0 \,.$$

Hence $\tilde{a}|C = f$, and clearly $\|\tilde{a}\| = \|f\|$. This completes the proof. □

We close this section by an example of Mokobodzki of non-unique representing measures *pseudo carried* by the extreme boundary of a simplex. (For the definition of the term "pseudo carried" see the Notes to Ch. I, § 4.) It also follows from this example that the set of boundary measures is *strictly* contained in the set of measures pseudo carried by the extreme boundary.

Proposition II. 3. 20. *There exists a simplex K with a point z such that*

(i) *$\partial_e K$ is a Borel set.*
(ii) $\mu_z(\partial_e K) = 1$.
(iii) *There exists a measure $\mu' \in M^+_z(K)$ such that $\mu'(\partial_e K) = 0$, $\mu'(C) = 0$ for every G_δ-set C disjoint from $\partial_e K$.*

Proof. Let Y be a compact Hausdorff space with a point y_0 which is not a G_δ, and let μ be a non-atomic measure in $M^+_1(Y)$. For example we can take y_0 to be any point in an uncountable product of unit intervals, and μ to be the corresponding product of Lebesgue measure by itself.

Let A be the space of all $f \in C(Y)$ satisfying $f(y_0) = \mu(f)$, let K be the state space of $(A, 1_Y)$ and let $\varphi: Y \to K$ be defined by $y \leadsto \tilde{y}$, where $\tilde{y}(a) = a(y)$ for all $a \in A$.

We observe first that we have the following implication similar to (3.32):

$$\varphi(y_1) = \varphi(y_2) \;\Rightarrow\; \exists \alpha \in \mathbb{R}: \varepsilon_{y_1} - \varepsilon_{y_2} = \alpha(\varepsilon_{y_0} - \mu) \,. \tag{3.40}$$

From this and the non-atomicity of μ we conclude that φ is a bijection of $Y\backslash\{y_0\}$ onto $\partial_e K$ by means of an argument similar to that of the preceding proof. We omit the details since no new ideas are involved.

Let $\varphi\mu \in M_1^+(K)$ be the direct image of μ by φ. It is easily verified that $\varphi\mu$ represents $\varphi(y_0)$ and that $\varphi\mu \neq \varepsilon_{\varphi(y_0)}$. Hence $\varphi(y_0)$ is not an extreme point of K. It follows that $\partial_e K = \varphi(Y)\backslash\{y_0\}$, and so $\partial_e K$ is a *Borel* set. It also follows that a measure $v \in M_1^+(K)$ is a boundary measure iff $v(\partial_e K)=1$. In fact, this condition is always sufficient since $\partial_e K$ is contained in every boundary set B_f with $f \in C_{\mathbb{R}}(K)$; but in the present situation it is also necessary for if $v(\partial_e K)<1$, where v is some boundary measure, then

$$v(\{y_0\}) = v(\overline{\partial_e K}) - v(\partial_e K) > 0,$$

and this is impossible since a boundary measure never has any *point* masses off the extreme boundary.

Now it only remains to prove the last statement (iii). We define $z=\varphi(y_0)$ and $\mu'=\varepsilon_z$. Clearly μ' represents z, and μ' is no boundary measure. We consider a G_δ-set C disjoint from $\partial_e K$, say $C=\bigcap_{n=1}^{\infty} U_n$ where $U_1, U_2, \dots$ are open subsets of K. Since φ maps $Y\backslash\{y_0\}$ onto $\partial_e K$, we shall have

$$\bigcap_{n=1}^{\infty} \varphi^{-1}(U_n) = \varphi^{-1}(C) \subset \{y_0\}. \tag{3.41}$$

From this we conclude that $\mu'(C)=0$, for otherwise $z \in C$, or equivalently $y_0 \in \varphi^{-1}(C)$, which would turn the inclusion of (3.41) into an equality, contrary to the hypothesis that y_0 was not a G_δ. The proof is complete.* □

Notes. The general concept of a simplex was introduced by Choquet in 1956 [100]. The original definition is somewhat different from the lattice-definition of the preceding section. It applies to an arbitrary convex subset K of an affine space H, and it declares K to be a *simplex* if the set of all homothets and translates of K is closed under finite intersections. To see the equivalence of the two definitions, one should *embed* H in a vectorspace E such that $0 \notin H$, and consider the ordering defined by the cone C with top-point 0 and base K. E is a lattice iff the intersection of any two translates of C is a new translate, and it can be shown that a relativization of this statement to H yields Choquet's original statement about homothets and translates. Some boundedness requirement is needed in the verification of this fact, and the least restrictive one is the "linear compactness" used by Kendall in 1962 [242]. The reader is referred to Kendall's paper for the details of the proof.

The Theorems II.3.7–II.3.8 were found by Choquet and Meyer in the early sixties [288], [104], [101], [111]. Theorem II.3.10 was proved in 1965 by Edwards [143], but the Corollary II.3.11 was also established independently by Lindenstrauss in 1964 [273] and by Semadeni in 1965 [361]. The importance of the Riesz decomposition property was first noted by F. Riesz in his fundamental paper of 1940 [342]. Note in this connection that the statement (3.3) can be considered as a special case of the lattice theoretic version of the Schreier Refinement Theorem, proved by Ø. Ore in 1935 [317] (cf. also [61]). In fact, since every vector lattice is distributive, and hence also modular, one can apply the refinement theorem to the chains $(0, u_1, u_1+u_2, \ldots, u)$ and $(0, v_1, v_1+v_2, \ldots, u)$. The differences between the final and initial points of the pairwise equivalent order intervals (or "formal quotients") of the two refined chains are pairwise equal, and these differences are the desired elements w_{ij}. It is also worth noting that although the Riesz decomposition property holds for the (self adjoint part of) a C^*-algebra iff it is commutative (=lattice ordered), there still exists an "asymmetric Riesz decomposition" in the general case, as was proved recently by G. Kjærgaard Pedersen [318].

Theorem II.3.12 is essentially due to Choquet [109] (cf. also [160]), and the non-metrizable counterexample of Proposition II.3.19 seems to be new, but the method of construction is of course standard and goes back to the fundamental paper of Bishop and de Leeuw from 1959 [63]. The example in Proposition II.3.20 was given by Mokobodzki in 1962 [297].

Clearly the simplexes of Proposition II.3.17 and Proposition II.3.19 both have non-closed extreme boundary. However, the constructions are unnecessarily complicated for the purpose of providing a simplex K with non-closed $\partial_e K$. The simplest example to this effect is probably one which can be obtained from the simplex $M_1^+(\bar{N})$ where $\bar{N}$ is the one-point compactification of N, after a deformation into a new simplex in which ε_∞ is identified with the mid-point of the segment $[\varepsilon_1, \varepsilon_2]$. Equivalently one could start with the positive part of the unit ball of l_1 endowed with the w^*-topology; here the origin plays the role of the point ε_∞ in the former realization. It should be mentioned that this example is of some interest in other respects as well; e.g. it shows that the closure of a face of a simplex may no longer be a face. This example was presented independently by Alfsen [5] and Lindenstrauss [273] in 1964. The reader is referred to these two papers for details.

In the above examples $\overline{\partial_e K}\backslash\partial_e K$ consists of just one point. However, much more can be achieved. Already in 1960 Thue Poulsen gave an example of a simplex K such that $\partial_e K$ is dense in K [338].

§ 4. Bauer Simplexes and the Dirichlet Problem of the Extreme Boundary

A (compact) simplex K with closed extreme boundary $\partial_e K$ is called a *Bauer simplex.*

Theorem II.4.1. *(Bauer) If K is a compact convex set then the following statements are equivalent:*

(i) *K is a Bauer simplex.*

(ii) *Every point in K is the barycenter of a unique positive and normalized measure supported by $\overline{\partial_e K}$.*

(iii) *K is a simplex and $x \rightsquigarrow \mu_x$ is a continuous map from K into $M_1^+(K)$ endowed with the vague topology.*

(iv) *$\hat{f} \in A(K)$ for all $f \in P(K)$.*

(v) *$A(K)$ is a lattice in the natural ordering of functions.*

Proof. (i)$\Rightarrow$(ii) If K is a Bauer simplex and if $x \in K$, $\mu \in M_x^+(K)$ and $\operatorname{Supp}(\mu) \subset \overline{\partial_e K}$, then

$$\mu(B_f) \geqq \mu(\partial_e K) = \mu(\overline{\partial_e K}) = 1$$

for all $f \in C_{\mathbb{R}}(K)$. Hence μ is a boundary measure, and so $\mu = \mu_x$.

(ii)$\Rightarrow$(iii) It follows from the statement (ii) and from the general fact that every boundary measure is supported by $\overline{\partial_e K}$ (Prop. I.4.6), that there can not be more than one positive and normalized boundary measure representing a given point in K. Hence K is a simplex. Also it follows that $\mu \rightsquigarrow x_\mu$ is a continuous $1-1$ map of the set of positive normalized measures on $\overline{\partial_e K}$ onto K. The domain of this map is compact and the inverse map $x \rightsquigarrow \mu_x$ is continuous.

(iii)$\Rightarrow$(iv) If (iii) is satisfied and $f \in P(K)$, then $\hat{f}$ is affine by Theorem II.3.7. Using formula (3.6) we can decompose $\hat{f}$ as follows

$$\hat{f}: x \rightsquigarrow \mu_x \rightsquigarrow \mu_x(f). \tag{4.1}$$

Form this the continuity follows.

(iv)$\Rightarrow$(v) If (iv) is satisfied and $a_1, a_2 \in A(K)$, then $\widehat{a_1 \vee a_2} \in A(K)$. By the definition of upper envelopes, $\widehat{a_1 \vee a_2}$ is the l.u.b. of a_1 and a_2 in $A(K)$. Hence $A(K)$ is a lattice.

(v)$\Rightarrow$(i) We assume $A(K)$ to be a lattice and consider functions $a_1, \ldots, a_n \in A(K)$ with a l.u.b. a in $A(K)$. By the definition of upper envelopes $a = \widehat{a_1 \vee \cdots \vee a_n}$. In particular $\widehat{a_1 \vee \cdots \vee a_n}$ is affine, and it follows from the Remark to Theorem II.3.7 that K is a simplex.

It remains to be proved that $\partial_e K$ is closed. In this connection we claim that in the formula

$$\partial_e K = \bigcap \{B_f \mid f \in C_{\mathbb{R}}(K)\}$$

one may replace $C_{\mathbb{R}}(K)$ by the cone of all functions $a_1 \vee \cdots \vee a_n$ where $a_i \in A(K)$ for $i=1,\ldots,n$. In fact let $x \notin \partial_e K$, say $x=\frac{1}{2}y+\frac{1}{2}z$, and let b_1, b_2 be two functions in $A(K)$ such that

$$b_1(y)=b_2(z)=1, \qquad b_1(z)=b_2(y)=0\,.$$

(Such functions exist by the Hahn-Banach Theorem.) Now

$$[b_1 \vee b_2](x)=\tfrac{1}{2}, \qquad \widehat{b_1 \vee b_2}(x) \geqq 1\,,$$

and the claim is proved.

If $f=a_1 \vee \cdots \vee a_n$ where $a_i \in A(K)$ for $i=1,\ldots,n$, then $\hat{f}$ is continuous by hypothesis, and hence B_f is closed. It follows that $\partial_e K$ is closed, and the proof is complete. □

We observe that each of the statements (ii), (iv), (v) of Theorem II.4.1 is a strengthened counterpart of a corresponding property for general simplexes. Thus, (ii) is stronger than the uniqueness statement of Theorem II.3.6, (iv) is stronger than the statement (iv) of Theorem II.3.7, and (v) is stronger than the Riesz interpolation property (Cor. II.3.11). Statement (iii) of Theorem II.4.1 is stronger than the measurability statement of Corollary II.3.15, and this result is particularly interesting because it establishes a $1-1$ correspondence between Bauer simplexes and compact Hausdorff spaces:

Corollary II.4.2. *If X is an arbitrary compact Hausdorff space then $M_1^+(X)$ is a Bauer simplex, and $\partial_e M_1^+(X) \cong X$. Conversely, every Bauer simplex K can be obtained in this way: in fact K is homeomorphic and affinely isomorphic to $M_1^+(\partial_e K)$.*

Proof. 1. For every compact Hausdorff space X, $M_{\mathbb{R}}(X)$ is a vector lattice, and hence $M_1^+(X)$ is a Bauer simplex (regularly embedded in $M_{\mathbb{R}}(X)$). Also $\partial_e M_1^+(X) \cong X$ (cf. e.g. [140]), and so $\partial_e M_1^+(X)$ is compact. Hence $M_1^+(X)$ is a Bauer simplex with the desired property.

2. For every Bauer simplex K the map $x \rightsquigarrow \mu_x$ defines a homeomorphism and an affine isomorphism of K onto $M_1^+(\overline{\partial_e K})=M_1^+(\partial_e K)$ (cf. formula (4.1)). This completes the proof. □

The next theorem contains Proposition II.3.13 as a proper subset.

Theorem II.4.3. *(Bauer) If K is a compact convex set, then the following statements are equivalent:*

(i) *K is a Bauer simplex.*

(ii) *Every continuous function $f\colon \partial_e K \to \mathbb{R}$ can be extended to a function in $A(K)$.*

(iii) *Every bounded continuous function $f\colon \partial_e K \to \mathbb{R}$ can be extended to a function in $A(K)$.*

(iv) *Every uniformly continuous function $f\colon \partial_e K \to \mathbb{R}$ can be extended to a function in $A(K)$.*

Moreover, if K is a Bauer simplex, then the extension of (ii) *can be made norm preserving.*

Proof. (i) $\Rightarrow$ (ii) Application of Proposition II.3.13.

(ii) $\Rightarrow$ (iii) Trivial.

(iii) $\Rightarrow$ (iv) Trivial since every uniformly continuous function on $\partial_e K$ is bounded. (The uniformity on $\partial_2 K$ is totally bounded.)

(iv) $\Rightarrow$ (i) We assume (iv) and consider two functions $a_1, a_2 \in A(K)$. Let $a \in A(K)$ be an extension of $a_1 \vee a_2 | \partial_e K$. Thus a and $a_1 \vee a_2$ coincide on $\partial_e K$ and it follows that a is the l.u.b. of a_1 and a_2 in $A(K)$. By Theorem II.4.1 K is a Bauer simplex.

The last statement of the theorem follows from Proposition II.3.13. □

The problem on the extension of a continuous real valued function on $\partial_e K$ to a function in $A(K)$ is often referred to as the "Dirichlet problem of the extreme boundary". By Theorem II.4.3 it is solvable for *every* continuous function on $\partial_e K$ iff K is a Bauer simplex. If K is an arbitrary compact convex set, then there are certain obvious necessary conditions which f must satisfy to be extendable within $A(K)$. One such condition is uniform continuity; another is $\mu(\bar{f}) = \nu(\bar{f})$, where μ, ν are two measures in $Z_1^+(K)$ with common barycenter and $\bar{f}$ is the continuous extension of f to $\overline{\partial_e K}$. (Note that $\mu(\bar{f})$, $\nu(\bar{f})$ are well defined since μ, ν are concentrated within $\overline{\partial_e K}$. Cf. Prop. I.4.6.) These conditions are not sufficient. On the extreme boundary of a simplex with non-closed extreme boundary there must exist a uniformly continuous function which is not extendable to a function in $A(K)$, although the above compatibility with representing boundary measures is vacuously satisfied.

We shall now establish a necessary and sufficient condition for the solvability of the Dirichlet problem of the extreme boundary, and we shall find it convenient to reformulate the above compatibility requirement by use of affine dependences. We recall that an *affine dependence* on a compact convex set K is a (signed) measure μ such that $\mu(a)=0$ for every $a\in A(K)$. (Cf. Ch. I, § 2.)

If μ is an affine dependence on K, then the Jordan decomposition of μ yields an equation

$$\mu=\alpha_1\mu_1-\alpha_2\mu_2; \quad \alpha_1=\alpha_2>0, \tag{4.2}$$

where $\mu_1, \mu_2\in M_1^+(K)$, and where μ_1 and μ_2 have common barycenter. Conversely, if μ_1, μ_2 are two measures in $M_1^+(K)$ with common barycenter, then every measure of the form (4.2) is an affine dependence.

An affine dependence μ on K is said to be a *boundary affine dependence* if μ is a boundary measure. Note in this connection that a signed measure μ is a boundary measure iff the Jordan components are. (Cf. e.g. formula (4.10) of Ch. I.) Note also that Choquet's uniqueness theorem (Th. II.3.6) can be restated in the following form: *A compact convex set K is a simplex iff it does not admit anv non-zero boundary affine dependences.*

In the remaining part of this section we shall make extensive use of the results and notation of Chapter I, § 5.

Lemma II.4.4. *Let K be a compact convex set, let $f:\partial_e K\to\mathbb{R}$, and assume that $\hat{f}$ and $\check{f}$ coincide on the set $X=\overline{\partial_e K}$. Let $A\subset C_{\mathbb{R}}(X)$ be the space of all functions $a|X$ where $a\in A(K)$, and let B be the linear span of A and $\bar{f}$, where $\bar{f}$ is the common restriction of $\hat{f}$ and $\check{f}$ to X. Then A and B determine the same envelopes, i.e.*

$$\hat{g}_A=\hat{g}_B, \quad \text{for all } g\in C_{\mathbb{R}}(X). \tag{4.3}$$

Proof. Clearly $\hat{g}_B\leqq\hat{g}_A$ since $A\subset B$. To prove the converse relation we consider a point $x\in X$ and an arbitrary $\varepsilon>0$. Now there is an $a_0\in A(K)$ and a $\lambda\in\mathbb{R}$ such that

$$g\leqq a_0|X+\lambda\bar{f}, \quad a_0(x)+\lambda\bar{f}(x)<\hat{g}_B(x)+\frac{\varepsilon}{2}. \tag{4.4}$$

At this point we have to study three different cases. If $\lambda>0$, then we choose $a_1\in A(K)$ such that

$$\bar{f}=\hat{f}|X\leqq a_1|X, \quad a_1(x)<\bar{f}(x)+\frac{\varepsilon}{2\lambda}.$$

If $\lambda<0$, then we choose $a_1\in A(K)$ such that

$$a_1|X\leqq\check{f}|X=\bar{f}, \quad \bar{f}(x)+\frac{\varepsilon}{2\lambda}<a_1(x).$$

If $\lambda=0$, then the choice of $a_1 \in A(K)$ is immaterial. In any case we obtain

$$g \leqq (a_0+\lambda a_1)|X, \qquad a_0(x)+\lambda a_1(x)<\hat{g}_B(x)+\varepsilon. \tag{4.5}$$

Since $\varepsilon>0$ was arbitrary, this means that $\hat{g}_A(x) \leqq \hat{g}_B(x)$ and the lemma is proved. □

Theorem II.4.5. *Let K be a compact convex set and $f: \partial_e K \to \mathbb{R}$ a continuous function. Then f can be extended to a function in $A(K)$ iff the following two conditions are satisfied:*

(i) The envelopes $\hat{f}$ and $\check{f}$ coincide on $\overline{\partial_e K}$.

(ii) The common restriction of $\hat{f}$ and $\check{f}$ to $\overline{\partial_e K}$ is annihilated by every boundary affine dependence.

Proof. 1. The necessity is obvious, for if $a \in A(K)$ and $a|\partial_e K=f$, then $\hat{f}=\check{f}=a$ and $v(\hat{f})=v(\check{f})=v(a)=0$ for every affine dependence v.

2. To prove the sufficiency we assume (i) and (ii), and we adopt the notation of the preceding lemma.

We shall be through if we can prove $\bar{f} \in A$, and for this it suffices (by a standard duality argument based on the Hahn-Banach Theorem), to establish the implication:

$$\mu \in M_{\mathbb{R}}(X), \quad \mu(a)=0 \quad \text{all } a \in A \;\Rightarrow\; \mu(\bar{f})=0. \tag{4.6}$$

We consider a measure $\mu \in M_{\mathbb{R}}(X)$ such that $\mu(a)=0$ for all $a \in A$, and we decompose it into $\mu=\alpha_1\mu_1-\alpha_2\mu_2$ where $\alpha_1=\alpha_2>0$ and $\mu_1, \mu_2 \in M_1^+(X)$. We denote by $\tilde{A}$ and $\tilde{B}$ the max-stable cones generated by A and B, respectively, and we apply Theorem I.5.19 to construct two $\tilde{B}$-maximal measures $v_1, v_2 \in M_1^+(X)$ such that $\mu_i \prec_{\tilde{B}} v_i$ for $i=1,2$. For every $k \in B$ we shall have $\mu_i(k) \leqq v_i(k)$ and also $\mu_i(-k) \leqq v_i(-k)$; hence μ_i and v_i coincide on B for $i=1,2$.

Thus

$$v_i(a)=\mu_i(a), \quad \text{all } a \in A, \qquad i=1,2, \tag{4.7}$$

and

$$v_i(\bar{f})=\mu_i(\bar{f}), \qquad i=1,2. \tag{4.8}$$

By Proposition I.5.9 and Lemma II.4.4 the two max-stable cones $\tilde{A}$ and $\tilde{B}$ determine the same maximal measures. Hence the measures v_1, v_2 are also $\tilde{A}$-maximal.

The cone $\tilde{A}$ is uniformly dense in the cone of all restrictions to X of continuous convex functions on K. (Cor. I.1.3.) Hence the ordering $\prec_{\tilde{A}}$, defined by $\tilde{A}$, is identical with the restriction of Choquet's ordering from $M_1^+(K)$ to $M_1^+(X)$. (By common abuse of language every measure on X is also considered as a measure on K, and every measure on K with support in X is considered as a measure on X.) It follows that the $\tilde{A}$-maximal measures in $M_1^+(X)$ are *boundary measures* on K. In particular, ν_1, ν_2 are boundary measures, and it follows from (4.7) that $\nu = \alpha_1 \nu_1 - \alpha_2 \nu_2$ is a *boundary affine dependence*. By the hypothesis (ii) we shall have $\nu(\bar{f}) = 0$, and now it follows by (4.8) that $\mu(\bar{f}) = \nu(\bar{f}) = 0$. This completes the proof of the implication (4.6). □

Remark. Generally $\hat{f}$ and $\check{f}$ coincide on $\partial_e K$. (Cor. I.4.2.) Hence condition (i) is redundant when $\partial_e K$ is closed. Note also that the notion of a "boundary affine dependence" is (much) more restrictive than that of an affine dependence supported by $\overline{\partial_e K}$. For example, if K is a simplex in which $\partial_e K$ is dense, then there are no non-zero boundary affine dependences, but there are infinitely many affine dependences supported by $K = \overline{\partial_e K}$. If $f \in C_{\mathbb{R}}(\overline{\partial_e K})$, and if f is annihilated by every affine dependence on $\overline{\partial_e K}$, then we can extend f to a function in $A(K)$. This, however, is no deep result: it follows directly by use of implication (4.6).

Corollary II.4.6. *Let K_1, K_2 be compact convex sets, let φ be a continuous map of $\partial_e K_1$ into K_2, and assume that $\widehat{a \circ \varphi}$ and $\widecheck{a \circ \varphi}$ coincide on $\partial_e K_1$ for all $a \in A(K_2)$. Then φ is uniformly continuous with a continuous extension $\bar{\varphi}: \overline{\partial_e K_1} \to K_2$; and if the direct image $\bar{\varphi}\nu$ is an affine dependence on K_2 for every boundary affine dependence ν on K_1, then φ can be extended to a continuous affine map of K_1 into K_2.*

Proof. For every $a \in A(K_2)$ the common restriction of $\widehat{a \circ \varphi}$ and $\widecheck{a \circ \varphi}$ to $\partial_e K_1$ is continuous, hence uniformly continuous. It follows that the function

$$a \circ \varphi = \widehat{a \circ \varphi} | \partial_e K_1 = \widecheck{a \circ \varphi} | \partial_e K_1$$

is uniformly continuous on $\partial_e K_1$ for every $a \in A(K_2)$. By compactness, the weak topology on K_2 (defined by $A(K_2)$) coincides with the given topology. Hence φ is a uniformly continuous map of $\partial_e K_1$ into K_2.

Again let $a \in A(K_2)$. The common restriction of $\widehat{a \circ \varphi}$ and $\widecheck{a \circ \varphi}$ to $\overline{\partial_e K_1}$ is just $a \circ \bar{\varphi}$; hence we obtain for every boundary affine dependence ν on K_1

$$\int (a \circ \bar{\varphi}) d\nu = \int a \, d(\bar{\varphi}\nu) = 0 . \tag{4.9}$$

By Theorem II.4.5 there is an $a' \in A(K_1)$ such that $a' | \overline{\partial_e K_1} = a \circ \bar{\varphi}$. Thus we have proved that $a \circ \varphi$ can be extended to a function $a' \in A(K_1)$ for every $a \in A(K_2)$.

For every $x \in K_1$ let μ be some boundary measures representing x, and let z be the barycenter of $\bar{\varphi}\mu$. The point $z \in K_2$ is independent of the choice of μ; for if μ' is some other boundary measure representing x, then $\mu - \mu'$ is a boundary affine dependence on K_1, hence $\bar{\varphi}\mu - \bar{\varphi}\mu'$

is a boundary affine dependence on K_2, and then the barycenter of $\overline{\varphi}\mu'$ is the same as that of $\overline{\varphi}\mu$. Clearly the map $\tilde{\varphi}: x \rightsquigarrow z$ is affine; and it extends φ, for if $x \in \partial_e K$ then ε_x is the only measure representing x.

It remains to be proved that $\tilde{\varphi}$ is continuous. With the above notation we obtain for an arbitrary $a \in A(K_2)$ and an arbitrary $x \in K_1$:

$$\begin{aligned} a'(x) &= \int a'\,d\mu = \int (a \circ \overline{\varphi})\,d\mu \\ &= \int a\,d(\overline{\varphi}\mu) = a(z) = a(\tilde{\varphi}(x)). \end{aligned}$$

Hence $a' = a \circ \tilde{\varphi}$, and it follows that $\tilde{\varphi}$ is a continuous map from the given topology of K_1 to the weak topology of K_2. Again the weak topology of K_2 will coincide with the given topology, and the proof is complete. □

Remark. Metrizability will simplify (ii) of Th. II.5.4 and the corresponding condition of Cor. II.4.5.

Notes. Theorems II.4.1–II.4.2 were proved by Bauer in 1961 [40] Theorem II 4.5 was proved for metrizable compact convex sets by Alfsen in 1966 [8]; then for arbitrary (compact) simplexes by Effros and Lazar in 1967–68 [156], [260]; and finally for arbitrary compact convex sets by Alfsen in 1968 [11]. The proof of this theorem has been greatly simplified by J. E. Björk (unpublished), and it is his proof which is presented in this book.

§ 5. Order Ideals, Faces, and Parts

Throughout this section we shall consider a fixed compact convex subset K of a locally convex (Hausdorff) space E, and we shall study the relationship between faces of K and order ideals of $A(K)$. We recall that every (norm-) complete order-unit space (A, e) can be realized as an $A(K)$ with $K = S(A, e)$, and therefore the results will apply to complete order-unit spaces in general.

Without lack of generality we assume K regularly embedded in E (cf. Ch. II, § 2). Then $A(K) \cong E^*$; specifically every $a \in A(K)$ admits a unique (norm-preserving) extension $\overline{a} \in E^*$ (Prop. II. 2.2). Also $E \cong A(K)^*$ under the isomorphism $x \rightsquigarrow q_x$ (cf. formula (2.4)), and we shall have

$$q_x(a) = \overline{a}(x). \tag{5.1}$$

By definition $x \rightsquigarrow q_x$ is bicontinuous from the given topology of E to the w^*-topology of $A(K)^*$. Also it is seen to be an isometry from the base-norm of (E, K) to the customary norm of the Banach dual space

$A(K)^*$. In the sequel we shall refer to this base-norm as the *norm of* E, and we shall refer to the corresponding ordering (with positive cone $\tilde{K}$) as *the ordering of* E.

For every $G \subset K$ we shall write:

$$G^{\perp} = \{a \in A(K) \mid a(x) = 0 \text{ for all } x \in G\} \tag{5.2}$$

and for every $B \subset A(K)$:

$$B^{\perp} = \{x \in K \mid a(x) = 0 \text{ for } a \in B\}, \tag{5.3}$$

$$B^{0} = \{x \in E \mid \bar{a}(x) = 0 \text{ for all } a \in B\}. \tag{5.4}$$

Proposition II.5.1. *For every* $G \subset K$ *the set* $G^{\perp}$ *is a (norm-) closed order ideal of* $A(K)$. *For every* $B \subset A(K)$ *the set* $B^{\perp}$ *is a closed convex subset of* K, *and* B^0 *is a closed vector subspace of* E *containing* $B^{\perp}$. *If* B *is positively generated (i.e. if* $B \subset B^+ - B^+$*), then* $B^{\perp}$ *is a face of* K.

The *proof* is a straightforward verification. □

Remark. The requirement $B \subset B^+ - B^+$ is not necessary in the above proposition. It can be weakened to a requirement that B be "almost positively generated" (*perfect* in Bonsall's terminology [79]). Specifically, the annihilator $B^{\perp}$ of a vector subspace B of an order-unit space (A, e) is a face of the state space of (A, e) iff, for each $b \in B$ and $\varepsilon > 0$, there exists a $c \in B$ such that

$$-c - \varepsilon e \leq b \leq c + \varepsilon e. \tag{5.5}$$

This theorem is due to Ellis, and the reader is referred to his papers [164], [170] for details.

Proposition II. 5.2. *If* J *is an order ideal of* $A(K)$ *and* $A(K)/J$ *is Archimedean ordered, then* $J = J^{\perp\perp}$.

Proof. Trivially $J \subset J^{\perp\perp}$. To prove the reverse inclusion we consider an $a \notin J$. By the hypothesis on Archimedicity, $(A(K)/J, \varphi(1_K))$ is an order-unit space, and by Proposition II.1.7 it admits a state q such that $q(\varphi(a)) \neq 0$. Now $q \circ \varphi$ is a state on $(A(K), 1_K)$ which vanishes on J. Hence $q \circ \varphi \in J^{\perp}$, and so $a \notin J^{\perp\perp}$. This completes the proof. □

Remark. The requirement to Archimedicity is not necessary in the above proposition. It can be weakened to a requirement that $A(K)/J$

be "almost Archimedean". Specifically $J=J^{\perp\perp}$ iff J is an order ideal with canonical map $\varphi: A(K)\to A(K)/J$, such that

$$-\lambda\varphi(b)\leqq\varphi(a)\leqq\lambda\varphi(b)$$

for some $b\geqq 0$ and all $\lambda>0$ implies $\varphi(a)=0$. This theorem is also due to Ellis, and again the reader is referred to his papers [164], [169] for details.

If $F\subset K$, then $F^{\perp\perp}$ is the set of all points $x\in K$ for which

$$a_1, a_2\in A(K), \quad a_1|F=a_2|F \;\Rightarrow\; a_1(x)=a_2(x). \tag{5.6}$$

This set is sometimes referred to as F's "set of determinacy", and F is said to be *self-determining* if $F=F^{\perp\perp}$.

Proposition II.5.3. *If $F\subset K$, then $F^{\perp\perp}=L\cap K$ where L is the set $(F^{\perp})^0$, which in turn is the* (w*-) *closed linear span of F in $E(\cong A(K)^*)$.*

Proof. Straightforward application of the Hahn-Banach theorem. □

Observe that the faces of K are exactly the sets $K\cap M$ where M is an affine subspace of E which "supports" K, in the sense that $M\cap K\neq\emptyset$ and $K\setminus M$ is convex. (Since K is assumed to be located on a hyperplane H where $0\notin H$, we can as well assume M to be a linear subspace, i.e. $0\in M$.) By Proposition II.3.5 the self-determining faces of K are those closed faces $F=K\cap M$ for which the supporting subspace M can also be chosen to be closed.

It follows from the above remarks that *a face F is self-determining if the linear span of F in E is closed.* In particular, *every finite dimensional face is self-determining.* In the "starred" section at the end of this paragraph we shall give an example of a closed face F which is self-determining (and enjoys various other nice properties) although lin F is not closed. In fact, it is rather difficult to exhibit a closed face which is not self-determining. An example to this effect has been given by Asimow. It is presented in Ellis' lecture notes on affine functions and faces of convex sets [170].

A (partially) ordered vector space is said to be *boundedly positively generated* if there exists a $\lambda\in\mathbb{R}^+$ such that every element x of the space admits a decomposition

$$x=x_1-x_2; \quad \|x_1\|+\|x_2\|\leqq\lambda\|x\|; \quad x_1, x_2\geqq 0. \tag{5.7}$$

Lemma II. 5.4. *Let F be a closed convex subset of K, and $\|x\|_F$ denote the norm of an element x of the base-norm space* $(\operatorname{lin} F, F)$ *while $\|x\|$ denotes the norm induced from E. For every $x \in \operatorname{lin} F$ we have $\|x\| \leqq \|x\|_F$, and the ratio $\|x\|_F/\|x\|$ is the least possible value of the scalar λ in a decomposition of type* (5.7) *with $x_1, x_2 \in (\operatorname{lin} F)^+$.*

Proof. By definition, the norm of E is a base-norm; in fact it is the norm of the base-norm space (E, K). Since $K \supset F$, we immediately obtain that $\|x\| \leqq \|x\|_F$.

Moreover:

$$\begin{aligned}\|x\|_F &= \inf\{\lambda > 0 \mid x \in \lambda \operatorname{conv}(F \cup -F)\}\\ &= \inf\{\lambda > 0 \mid x = \lambda\lambda_1 y_1 - \lambda\lambda_2 y_2 ; \lambda_1 + \lambda_2 = 1 ; \lambda_1, \lambda_2 \geqq 0 ; y_1, y_2 \in F\}.\end{aligned}$$

and the infimum-value is effectively attained by virtue of the compactness of conv $(F \cup -F)$.

Writing $x_i = \lambda\lambda_i y_i$ for $i = 1,2$, we have $\|x_1\| + \|x_2\| = \lambda$. Hence:

$$\|x\|_F = \inf\{\|x_1\| + \|x_2\| \mid x = x_1 - x_2 ; x_1, x_2 \in (\operatorname{lin} F)^+\}, \tag{5.8}$$

and the proof is complete. □

Theorem II.5.5. *(Edwards) If F is a closed convex subset of K, then the following statements are equivalent:*

(i) $\operatorname{lin} F$ *is boundedly positively generated in the ordering and norm induced from E.*

(ii) $\operatorname{lin} F$ *is a* (w*-) *closed subspace of E* $(\cong A(K)^*)$.

(iii) $\operatorname{lin} F$ *is a norm-closed subspace of E.*

Proof. (i) ⇒ (ii) By lemma II.5.4 there is a $\lambda \in \mathbb{R}^+$ such that

$$\|x\| \leqq \|x\|_F \leqq \lambda \|x\|, \quad \text{all } x \in \operatorname{lin} F. \tag{5.9}$$

Now (ii) follows by application of the Krein-Šmulian Theorem in the form of Proposition II.2.2, since the set

$$\operatorname{lin} F \cap \operatorname{conv}(K \cup -K) = [\lambda \operatorname{conv}(F \cup -F)] \cap \operatorname{conv}(K \cap -K)$$

is compact.

(ii) ⇒ (iii) Trivial.

(iii) ⇒ (i) By Proposition II.1.12 $\operatorname{lin} F$ is complete in the base-norm $\|-\|_F$, and by the Open Mapping Theorem there exists a $\lambda \in \mathbb{R}^+$ such that (5.9) holds. Now (i) follows by Lemma II.5.4. □

Corollary II.5.6. *A closed face F of K is self-determining if* lin F *is boundedly positively generated.*

A closed convex subset F of K is said to have the *extension property* (or to be an "interpolation set") if every $a \in A(F)$ can be extended to an $\tilde{a} \in A(K)$; or briefly if $A(F;K) = A(K)$. Also we shall say that F has the *bounded extension property* if there is a $\lambda \in \mathbb{R}^+$ such that each $a \in A(F)$ admits an extension $\tilde{a} \in A(K)$ with $\|\tilde{a}\| \leqq \lambda \|a\|$. The latter notion will only be used temporarily, as it will turn out to be a consequence of the former.

Unless otherwise stated, we shall assume that $A(F)$ is endowed with the uniform norm $\|a\|_F$ and that $A(F;K)$ is endowed with the "extension norm":

$$\|a\|_q = \inf\{\|\tilde{a}\| \mid \tilde{a} \in A(K);\ \tilde{a}|F = a\}. \tag{5.10}$$

We denote the extension norm by the subscript "q" since it corresponds to the quotient norm of $A(K)/F^{\perp}$ under the natural isomorphism of this vector space and $A(F;K)$.

To every closed convex subset F of K we associate a *characteristic number* (or "extension coefficient") defined by

$$\rho_F = \sup\left\{\frac{\|a\|}{\|a\|_F} \;\middle|\; a \in A(F;K),\ a \neq 0\right\}. \tag{5.11}$$

Clearly $\rho_F < \infty$ if F has the bounded extension property.

Lemma II.5.7. *Let F be a closed convex subset of K, let* lin F *be endowed with the base-norm of* (lin F, F), *and let* $(F^{\perp})^0$ *be endowed with the norm induced from E. Then we have isometric isomorphisms:*

$$A(F)^* \cong \operatorname{lin} F, \qquad A(F;K)^* \cong (F^{\perp})^0, \tag{5.12}$$

and

$$\rho_F = \sup\left\{\frac{\|x\|_F}{\|x\|} \;\middle|\; x \in \operatorname{lin} F,\ x \neq 0\right\}. \tag{5.13}$$

Proof. 1. Observe first that the uniform norm of $A(F)$ is the order-unit norm of $(A(F), 1_F)$, and so the norm of $A(F)^*$ is the base-norm of $(A(F)^*, S)$ where S is the state space of $(A(F), 1_F)$.

For every $x \in \operatorname{lin} F$ we define $r_x \in A(F)^*$ by writing $r_x(a) = \bar{a}(x)$ for $a \in A(F)$. We note that $x \rightsquigarrow r_x$ is a $1-1$ map of F onto S, since $A(F)$ separates points of F and every state of $(A(F), 1_F)$ is a point state. Thus, r is a linear map from $(\operatorname{lin} F, F)$ to $(A(F)^*, S)$ which maps the base F biuniquely onto the base S; and it follows that r is an isometric isomorphism of the former space onto the latter.

2. We have already noted that we can identify $A(F;K)$ with $A(K)/F^{\perp}$, and it follows from elementary properties of normed linear spaces (cf. e.g. [86]) and from the identification $x \rightsquigarrow q_x$ of E and $A(K)^*$, that we have isometries:

$$\begin{aligned}(A(K)/F^{\perp})^* &\cong \{q \in A(K)^* \mid q(a)=0 \text{ all } a \in F^{\perp}\}\\ &\cong \{x \in E \mid q_x(a)=\bar{a}(x)=0 \text{ all } a \in F^{\perp}\}=(F^{\perp})^0.\end{aligned}$$

3. Using elementary properties of normed linear spaces (essentially the Hahn-Banach Theorem), one may replace the quotient of the two norms in (5.11) by the reciprocal quotient of the dual norms. By virtue of the isometries (5.12) just proved, this converts (5.11) into (5.13). □

By application of Lemma II.5.4 we now obtain the following:

Corollary II.5.8. *If F is a closed convex subset of K, then ρ_F is the least possible value of $\lambda \in [1,\infty]$ such that every $x \in \operatorname{lin} F$ admits a decomposition of the type* (5.7).

Theorem II.5.9. *If F is a closed convex subset of K, then the following statements are equivalent.*

(i) *F has the extension property.*

(ii) *F has the bounded extension property.*

(iii) $\rho_F < \infty$.

(iv) $\operatorname{lin} F$ *is boundedly positively generated in the ordering and norm induced from E.*

Proof. (i) ⇒ (ii) Generally the two spaces $A(F)$ and $A(F;K) \cong A(K)/F^{\perp}$ are complete in the respective norms. If $A(F)=A(F;K)$, then these two norms must be topologically equivalent by the Open Mapping Theorem. Hence there is a $\lambda \in \mathbb{R}^+$ such that $\|a\|_F \leqq \|a\|_q \leqq \lambda\|a\|_F$ for all $a \in A(F)$, and this means that F has the bounded extension property.

(ii) ⇒ (iii) Trivial.

(iii) ⇒ (iv) If $\rho_F < \infty$, then it follows from Lemma II.5.7 that the base norm of $(\operatorname{lin} F, F)$ is topologically equivalent to the norm induced from E. The space $\operatorname{lin} F$ is always complete in the base-norm by virtue of Proposition II.1.12; and in the present situation it must also be complete in the norm of E. Hence $\operatorname{lin} F$ is a norm-closed subspace of E, and statement (iv) follows from Theorem II.5.5.

(iv) ⇒ (i) If $\operatorname{lin} F$ is boundedly positively generated, then $\rho_F < \infty$ (Cor. II.5.8), and it follows by the definition of ρ_F that the "extension norm" (5.10) of $A(F;K) \cong A(K)/F^{\perp}$ is topologically equivalent to the uniform norm. Hence $A(F;K)$ is a complete, and hence closed, subspace of the space $A(F)$ with uniform norm. By Corollary I.1.5, $A(F;K)$ is uniformly dense in $A(F)$. Hence $A(F;K)=A(F)$, and the theorem is proved. □

Combining with Corollary II.5.6 we obtain the following:

Corollary II.5.10. *If F is a closed face of K and* $\operatorname{lin} F$ *is boundedly positively generated, then F is a self-determining face with the extension property.*

Remark. It is now clear that the properties listed in Theorem II.5.5 and Theorem II.5.9 are all mutually equivalent, and they are tantamount to $A(F)=A(F;K)$ with (topological) equivalence of the uniform norm of $A(F)$ and the "extension"- or "quotient"-norm of $A(F;K)\cong A(K)/F^{\perp}$ (cf. Ch. II. formula (5.11)). Note, however, that the spaces $A(F)$ and $A(K)/F^{\perp}$ may still be very different order theoretically. For example, if K is the compact convex set shown in the diagram of page 68, and if F is the intersection of K with the y-axis, then a function $a_0 \in A(F)$ with values 0,1 at the end-points of the line segment F will certainly be a positive element of $A(F)$, but it admits no positive extension in $A(K)$; hence the corresponding element of $A(K)/F^{\perp}$ is non-positive.

It is worth noting that in the above example, $A(K)/F^{\perp}$ is non-Archimedean. In fact the element of $A(K)/F^{\perp}$ which corresponds to $-a_0$, is seen to be majorized by every strictly positive multiple of the order unit (the equivalence class of 1_K); still it is non-negative.

Our next proposition shows that if J is an order ideal of $A(K)$ and $F=J^{\perp}$, then Archimedicity of $A(K)/J$ is a necessary and sufficient condition that the natural isomorphism of the vector spaces $A(K)/J$ and $A(F;K)$ be an *order*-isomorphism with respect to the quotient ordering of $A(K)/J$ and the pointwise ordering of $A(F;K)$.

Proposition II.5.11. *Let J be an order ideal of $A(K)$ and let $F=J^{\perp}$. Now $A(K)/J$ is Archimedean in the quotient ordering iff:*

$$\textit{for each } a_0 \in A(F;K)^+ \textit{ there exists a } c \in A(K)^+ \textit{ such that } c|F=a_0. \tag{5.14}$$

Proof. 1. We assume that $A(K)/J$ is Archimedean, and we observe that for each state p on the order-unit space $(A(K)/J, \varphi(1_K))$ the functional $p\circ\varphi$ is a state on $(A(K), 1_K)$; hence there is a point $x \in K$ such that

$$a(x)=[p\circ\varphi](a), \quad \text{for all } a \in A(K). \tag{5.15}$$

Moreover, $a(x)=p(\varphi(a))=0$ for all $a\in J$, and so $x\in J^{\perp}=F$.

Now we consider an $a_0\in A(F;K)^+$, and we choose a (not necessarily positive) element $b\in A(K)$ such that $b|F=a_0$. If p is an arbitrary state on $(A(K)/J,\varphi(1_K))$ and $x\in F$ is the corresponding point for which (5.15) holds, then

$$p(\varphi(b))=b(x)=a_0(x)\geqq 0\,.$$

It follows that $\varphi(b)\geqq 0$ (cf. Prop. II.1.7); and by the definition of quotient ordering there exists a $c\in A(K)^+$ such that $c\equiv b$ (mod. J). Now $c|F=b|F=a_0$, and formula (5.14) is proved.

2. Assuming (5.14) we shall prove $A(K)/J$ Archimedean. If

$$\varphi(a)\leqq\frac{1}{n}\varphi(1_K),\qquad n=1,2,\ldots,$$

then there exist elements $b_n\in J$ such that

$$a\leqq\frac{1}{n}1_K+b_n,\qquad n=1,2,\ldots.$$

Hence

$$a(x)\leqq\frac{1}{n},\quad\text{all } x\in F,\qquad n=1,2,\ldots;$$

and so $a|F\leqq 0$.

By (5.14) there is an element $b\in A(K)^+$ such that $b|F=-a|F$. Now $\varphi(a)=-\varphi(b)\leqq 0$, and the Archimedicity is proved. □

An order ideal $J\subset A(K)$ is said to be *Archimedean* if

(i) $A(K)/J$ is Archimedean in the quotient ordering,
(ii) J is positively generated;

and J is said to be *strongly Archimedean* if in addition

(iii) J^0 is positively generated.

It follows from Proposition II.1.6 that every Archimedean ideal of $A(K)$ is *closed* (in uniform norm), and also that J is the kernel of $\varphi:(A(K),1_K)\to(A(K)/J,\varphi(1_K))$. Conversely, if $\psi:(A(K),1_K)\to(A',e')$ is an order-homomorphism with kernel J into some order-unit space (A',e') then J is a positively generated order-ideal and $A(K)/J$ is order-isomorphic to $\psi(A)\subset A'$. Hence $A(K)/J$ is Archimedean.

Summarizing we have:

Proposition II.5.12. *An order ideal $J \subset A(K)$ is Archimedean iff it is the kernel of an order-homomorphism of $(A(K), 1_K)$ into some order-unit space.*

Theorem II.5.13. *If J is an Archimedean order-ideal of $A(K)$ and $F = J^{\perp}$, then F is a closed face of K with the property:*

$$\text{for each } a \in A(K) \text{ with } a|F \geqq 0 \text{ there exists a } c \in A(K)^+ \text{ such that } c \geqq a \text{ and } c|F = a|F; \tag{5.16}$$

and $J = F^{\perp}$. Conversely, if F is a closed face of K satisfying (5.16), *then $J = F^{\perp}$ is an Archimedean ideal of $A(K)$, and $F = J^{\perp}$.*

Proof.

1. Assume first that J is Archimedean. By definition J is positively generated, and so F is a closed face of K (cf. Prop. II.5.1).

To prove (5.16) we consider an $a \in A(K)$ with $a|F \geqq 0$. By Proposition II.5.12 there exists a $b_1 \in A(K)^+$ such that $b_1|F = a|F$. Hence $a - b_1 \in J$, and since J is positively generated we may choose $b_2 \in J^+$ so that $a - b_1 \leqq b_2$. Writing $c = b_1 + b_2$, we obtain $c \in A(K)^+ + J^+ \subset A(K)^+$ and $a \leqq c$. Since $b_2 \in J$ and $a - b_1 \in J$, we have $c|F = b_1|F = a|F$, and formula (5.16) is proved.

The equality $J = F^{\perp}$ follows by Proposition II.5.2.

2. Assume next that F is a closed face of K satisfying (5.16). Clearly the set $J = F^{\perp}$ is a closed order ideal of $A(K)$ (cf. Prop. II.5.1). Clearly, (5.16) implies (5.14), and so $A(K)/J$ is Archimedean. To prove J positively generated, we consider an $a \in J$. By (5.16) there is a $c \in A(K)^+$ such that $c \geqq a$ and $c|F = a|F = 0$. This gives the desired relation,

$$a = c - (c - a) \in J^+ - J^+ .$$

Again, the inclusion $F \subset J^{\perp}$ is trivial. To prove $J^{\perp} \subset F$, we assume that $x \in K \backslash F$. Since J is closed, we may apply the Hahn-Banach theorem to yield an $a \in A(K)$ such that

$$a(x) < 0 < a(y), \quad \text{all } y \in F .$$

By (5.16) there is a $c \in A(K)^+$ such that $a|F = c|F$. It follows that $a - c \in J$, whereas $[a-c](x) < 0$. Hence $x \notin J^\perp$, and the proof is complete. □

Motivated by Theorem II.5.13 we shall say that a face F of a compact convex set K is *Archimedean* if it is closed and has the property (5.16). Using this notion, we may restate the theorem as follows: *The map $J \rightsquigarrow F = J^\perp$ is an (order reversing) bijection of the Archimedean ideals of $A(K)$ onto the Archimedean faces of K; the inverse map being $F \rightsquigarrow F^\perp$.*

Theorem II.5.14. *An Archimedean order-ideal $J \subset A(K)$ is strongly Archimedean iff $F = J^\perp$ has the extension property, — or any other of the equivalent properties listed in Theorem* II.5.5 *and Theorem* II.5.9.

Proof. 1. If J is strongly Archimedean, then $J^0 = (J^0)^+ - (J^0)^+ = \operatorname{lin} F$, and it follows that $\operatorname{lin} F$ is a closed subspace of E. This completes the first part of the proof by statement (ii) of Theorem II.5.5.

2. If F has the extension property, then it has the bounded extension property by virtue of Theorem II.5.9; and this means that the vector spaces $A(F)$ and $A(F;K)$ coincide and have (topologically) equivalent norms. Passing to the dual spaces and making use of Lemma II.5.7 and Theorem II.5.13, we obtain $\operatorname{lin} F = (F^\perp)^0 = J^0$. Thus, J^0 is positively generated, and the proof is complete. □

Motivated by Theorem II.5.14 we shall say that a face F of a compact convex set K is *strongly Archimedean* if it is Archimedean and enjoys the extension property; and we note that the characteristic number ρ_F of a strongly Archimedean face F is *finite*, and that there exists, for every $a \in A(F)$ and every $\varepsilon > 0$, an $\tilde{a} \in A(K)$ such that:

$$\tilde{a}|F = a; \qquad \|\tilde{a}\| < (\rho_F + \varepsilon)\|a\| . \tag{5.17}$$

Also we may restate Theorem II.5.14 as follows: *The map $J \rightsquigarrow F = J^\perp$ is an (order-reversing) bijection of the strongly Archimedean ideals of $A(K)$ onto the strongly Archimedean faces of K; the inverse map being $F \rightsquigarrow F^\perp$.*

We are now in the position to prove the following counterpart of Proposition II.5.12:

Proposition II.5.15. *An order-ideal $J \subset A(K)$ is strongly Archimedean iff it is the kernel of a topological order-homomorphism of $(A(K), 1_K)$ into some order-unit space.*

Proof. 1. Assume first that J is strongly Archimedean and define $F=J^{\perp}$. It follows that the spaces $A(F;K)$ and $A(F)$ coincide, and that their norms are equivalent (cf. the preceding proof). We recall that the "extension norm" (5.10) of $A(F;K)$ corresponds to the quotient norm of $A(K)/J$ under the natural isomorphism of these two vector spaces; and we observe that the uniform norm of $A(F)=A(F;K)$ corresponds to the order-unit norm of $(A(K)/J, \varphi(1_K))$. Thus the two norms of $A(K)/J$ are equivalent, and this means that $\varphi:(A(K), 1_K)\to(A(K)/J, \varphi(1_K))$ is a topological order-homomorphism.

2. Assume next that there exists a topological order-homomorphism with kernel J of $(A(K), 1_K)$ into some order-unit space. Then the two norms of $A(K)/J$ are equivalent; hence the two norms of $A(F;K)$ are equivalent, (again with $F=J^{\perp}$). In particular, $A(F;K)$ is complete in uniform norm, and it follows from the density of $A(F;K)$ in $A(K)$ that these two spaces coincide. Thus, F has the extension property, and the proof is complete. □

A face F of K is said to be *exposed* if there exists an $a\in A(K)^+$ such that $a|F=0$ and $a(x)>0$ for all $x\in F\setminus K$; and it is said to be *relatively exposed* if there exists for every $x\in K\setminus F$, an $a\in A(K)^+$ (depending on x) such that $a|F=0$ and $a(x)>0$. A point x in K is said to be *(relatively) exposed* if $\{x\}$ is a (relatively) exposed face. Clearly, every (relatively) exposed point is an extreme point, but there are simple examples in $\mathbb{R}^2$ of extreme points which are not exposed or relatively exposed. Note that a relatively exposed face is *closed* and *self-determining.*

(Recall that we have assumed K regularly embedded in E. Without this assumption we might have $A(K)\neq A(K;E)$, and then the replacement of $A(K)^+$ by $A(K;E)^+$ in the above definition would yield a new and more restrictive notion of a (relatively) "E-exposed" face. This non-intrinsic concept will not be studied in the sequel.)

Proposition II.5.16. *A relatively exposed face F of K is exposed iff it is a G_δ.*

Proof. The necessity is obvious. To prove the sufficiency, we first observe that if F is a relatively exposed face and C is a compact subset of $K\setminus F$, then it follows by a simple compactness argument that there exists an $a\in A(K)^+$ such that $a|F=0$ and $a(x)>0$ for all $x\in C$. Then we assume F to be a relatively exposed G_δ-face, say

$$K\setminus F=\bigcup_{n=1}^{\infty} C_n,$$

where $\{C_n\}$ is a sequence of compact sets. Now there exist functions $a_n \in A(K)^+$ such that $a_n|F=0$, $a_n(x)>0$ for all $x \in C_n$, and $\|a_n\|<2^{-n}$ for $n=1,2,\ldots$. Writing $a=\sum_{n=1}^{\infty} a_n$, we obtain a function with the desired properties: $a \in A(K)^+$, $a|F=0$, and $a(x)>0$ for all $x \in K \setminus F$. □

Note that by the above proposition, the concepts of an *exposed* face and of a *relatively exposed* face will coalesce for *metrizable* compact convex sets.

Proposition II.5.17. *An Archimedean face $F \subset K$ is relatively exposed.*

Proof. For every $x \in K \setminus F$ there exists by Hahn-Banach separation a function $a \in A(K)$ such that $a|F \geqq 0$ and $a(x)<0$. By the definition of an Archimedean face there exists a $c \in A(K)^+$ such that $c \geqq a$ and $c|F=a|F$. Now $c-a$ belongs to $A(K)^+$, $(c-a)|F=0$, and $(c-a)(x)>0$. This completes the proof. □

Corollary II.5.18. *An Archimedean face of a metrizable compact convex set is exposed.*

The converse of Corollary II.5.18 is inexact. Every extreme point x of a circle K in $\mathbb{R}^2$ is exposed, but $\{x\}$ is not an Archimedean face of K, as can be seen directly from the definition (5.16).

Remark. We have previously studied the example shown in Fig. 1 on p. 68. If $x=(0,0)$, then the face $\{x\}$ of K is evidently non-Archimedean, since x is a non-exposed point. Dualizing, we observe that $J=\{x\}^{\perp}$ violates both requirements on Archimedean ideals: $A(K)/J$ is not Archimedean ordered, and J is not positively generated.

Passing over to the example of an extreme point x of a circle, we observe that $J=\{x\}^{\perp}$ will violate the second requirement only. Here $A(K)/J \cong \mathbb{R}$, but $J \neq J^+ - J^+$.

Theorem II.5.19. *Every closed face F of a simplex K is strongly Archimedean, and every $a \in A(F)$ can be extended with preservation of norm to an $\tilde{a} \in A(K)$. In particular $\rho_F=1$.*

Proof. 1. We shall first prove F Archimedean by establishing (5.16). To this end we consider an arbitrary $a \in A(K)$ such that $a|F \geqq 0$, and we define functions $f, g: K \to \mathbb{R}^+$ by writing

$$f(x) = \max(a(x), 0),$$
$$g(x) = \begin{cases} a(x) & \text{for } x \in F, \\ \|a\| & \text{for } x \in K \backslash F. \end{cases}$$

It is easily verified that f is continuous and convex, while g is l.s.c. and concave. Also $f \leqq g$, and by Edwards' theorem (Th. II.3.10) there exists a $c \in A(K)$ such that $f \leqq c \leqq g$. Clearly $c \in A(K)^+$, $c \geqq a$, and $c|F = a|F$.

2. It remains to prove that F has the norm-preserving extension property. To this end we consider an arbitrary $a \in A(F)$ and we define two functions $h, k: K \to \mathbb{R}$ as follows:

$$h(x) = \begin{cases} a(x) & \text{for } x \in F, \\ -\|a\| & \text{for } x \in K \backslash F; \end{cases}$$

$$k(x) = \begin{cases} a(x) & \text{for } x \in F, \\ \|a\| & \text{for } x \in K \backslash F, \end{cases}$$

and we observe that we can apply Edwards' theorem (Th. II.3.10) to yield an $\tilde{a} \in A(K)$ such that $h \leqq \tilde{a} \leqq k$. Now $\tilde{a}|F = a$ and $\|\tilde{a}\| = \|a\|$, and the proof is complete. □

Corollary II.5.20. *Every closed face of a simplex is relatively exposed. In particular, every closed face of a metrizable simplex is exposed.*

Specializing to one-point faces, we obtain:

Corollary II.5.21. *Every extreme point of a metrizable simplex is an exposed point.*

Let x be an arbitrary point of K. The smallest (not necessarily closed) face containing x, consists of all points y in K for which the line segment $[x, y]$ extends in K beyond the point x. In symbols:

$$y \in \text{face}(x) \Leftrightarrow \exists\ \varepsilon > 0: x + \varepsilon(x - y) \in K. \tag{5.18}$$

Observe that we can transform the formula

$$x+\varepsilon(x-y)\in K \tag{5.19}$$

into

$$y\in\alpha x-(\alpha-1)K, \tag{5.20}$$

by writing $1+\varepsilon^{-1}=\alpha$.

Hence if we write

$$D_\alpha(x)=(\alpha x-(\alpha-1)K)\cap K, \tag{5.21}$$

for $\alpha\geqq 1$, then we shall have

$$\text{face}(x)=\bigcup_{n=1}^{\infty} D_n(x). \tag{5.22}$$

Clearly, $D_\alpha(x)$ is compact for every $\alpha\geqq 1$, and so we have proved:

Proposition II.5.22. *If x is an arbitrary point of K, then face(x) is an F_σ.*

Two points $x,y\in K$ are said to be *equivalent* if the line segment $[x,y]$ extends in K beyond x and y.

In symbols:

$$x\sim y \Leftrightarrow \text{face}(x)=\text{face}(y). \tag{5.23}$$

The equivalence classes for this relation are called *parts*.

Clearly, the concept of a "part" is closely related to that of a "face", but we note that a part is never closed unless it reduces to a single point, which is then necessarily an extreme point.

For brevity we shall say that a segment $[x,y]\subset K$ "extends by ε beyond x" if (5.19) holds. If $[x,y]$ extends by ε beyond both x and y, then we shall simply say that $[x,y]$ "extends by ε".

Lemma II.5.23. *Let $x,y,z\in K$. If $[x,z]$ extends by α and $[y,z]$ extends by β, then $[x,y]$ extends by γ, where $(1+\gamma^{-1})=(1+\alpha^{-1})(1+\beta^{-1})$.*

Proof. Let γ be defined by the formula above, and define

$$\lambda=\frac{1+\beta}{1+\alpha+\beta},\qquad \mu=\frac{1+\alpha}{1+\alpha+\beta}.$$

It is easy to verify that

$$x+\gamma(x-y)=\lambda[x+\alpha(x-z)]+(1-\lambda)[z+\beta(z-y)],$$
$$y+\gamma(y-x)=\mu[y+\beta(y-z)]+(1-\mu)[z+\beta(z-x)].$$

Hence $x+\gamma(x-y)\in K$ and $y+\gamma(y-x)\in K$; and so $[x,y]$ extends by γ. □

It follows from (5.22) and (5.23) that

$$x\sim y \Leftrightarrow \exists\, \alpha\geqq 1: \quad y\in D_\alpha(x),\ x\in D_\alpha(y). \tag{5.24}$$

Hence for every part P of K we can define a function $\alpha_P: P\times P\to[1,\infty[$ as follows:

$$\alpha_P(x,y)=\inf\{\alpha\in[1,\infty[\ \mid y\in D_\alpha(x),\ x\in D_\alpha(y)\}. \tag{5.25}$$

By virtue of the equivalence of (5.19) and (5.20), we can also state the definition as follows:

$$\alpha_P(x,y)=\inf\{1+\varepsilon^{-1}\mid [x,y] \text{ extends by } \varepsilon\}. \tag{5.26}$$

Theorem II.5.24. *(Bear-Weiss) If P is a part of K, then the function α_P satisfies.*

(i) $\alpha_P(x,y)=1\Leftrightarrow x=y$,

(ii) $\alpha_P(x,y)=\alpha_P(y,x)$,

(iii) $\alpha_P(x,y)\leqq\alpha_P(x,z)\cdot\alpha_P(y,z)$.

Moreover, for any two points $x,y\in K$, the value $\alpha_P(x,y)$ is the least positive number α for which we have a Harnack type inequality.

(iv) $\dfrac{1}{\alpha}\leqq\dfrac{a(y)}{a(x)}\leqq\alpha$, all $a\in A(K)^+$.

Proof. The statements (i), (ii), (iii) follow easily from the definition (5.26) by use of Lemma II.5.23.

To prove the relation

$$\frac{1}{\alpha_P(x,y)}\leqq\frac{a(y)}{a(x)}\leqq\alpha_P(x,y), \quad \text{all } a\in A(K)^+, \tag{5.27}$$

we consider an arbitrary real number $\alpha>\alpha_P(x,y)$. By the definition (5.25), $y\in D_\alpha(x)$. Hence there exists a $z\in K$ such that

$$y=\alpha x-(\alpha-1)z. \tag{5.28}$$

Hence

$$a(y)=\alpha a(x)-(\alpha-1)a(z)\leqq\alpha a(x), \quad \text{all } a\in A(K)^+.$$

This gives the inequality at the right hand side of (5.27), since $\alpha > \alpha_P(x,y)$ was arbitrary. Interchanging x and y, we obtain the inequality at the left hand side.

Finally we assume $\alpha < \alpha_P(x,y)$. Now, either $y \notin D_\alpha(x)$ or $x \notin D_\alpha(y)$; and we may as well assume $y \notin D_\alpha(x)$. Then the point $z \in E$ defined by (5.28), is exterior to K. By Hahn-Banach separation, there exists a continuous affine function $\bar{a}$ on E such that $\bar{a}$ is strictly positive on K and $\bar{a}(z) < 0$.

Now

$$a(y) = \alpha a(x) - (\alpha - 1) a(z) > \alpha a(x),$$

and the function $a = \bar{a}|K$ will satisfy

$$\frac{a(y)}{a(x)} > \alpha, \qquad a \in A(K)^+. \tag{5.29}$$

This violates (iv), and the proof is complete. □

It follows from Theorem II.5.24 that $\delta_P(x,y) = \log \alpha_P(x,y)$ is a metric on a given part of K, and this metric is usually called the *part metric*.

We shall close this section by a measure theoretic characterization of the parts of a simplex. In this connection we shall use the brief notation $\nu \in L^\infty(\mu)$ to mean that $\nu \ll \mu$ and $\dfrac{d\nu}{d\mu} \in L^\infty(\mu)$.

Theorem II. 5.25 *Two points x, y of a simplex K belong to the same part iff $\mu_y \in L^\infty(\mu_x)$ and $\mu_x \in L^\infty(\mu_y)$.*

Proof. 1. If $x \sim y$, then there exist an $\alpha \geq 1$ and a $z \in K$ such that

$$y = \alpha x - (\alpha - 1) x.$$

Hence

$$\mu_y = \alpha \mu_x - (1 - \alpha) \mu_z \leqq \alpha \mu_x$$

It follows that $\mu_y \ll \mu_x$ and $\left\| \dfrac{d\mu_y}{d\mu_x} \right\|_\infty \leqq \alpha$. This proves $\mu_y \in L^\infty(\mu_x)$, and we obtain the opposite relation by interchanging x and y.

2. Conversely, assume $\mu_y \in L^\infty(\mu_x)$ and $\mu_x \in L^\infty(\mu_y)$. If $\left\| \dfrac{d\mu_y}{d\mu_x} \right\|_\infty \leqq \alpha$,

then $\mu_y \leqq \alpha \mu_x$. Now the measure

$$\nu = (1-\alpha)^{-1}(\alpha \mu_x - \mu_y),$$

will be a probability measure, and we denote its barycenter by z. Then we shall have

$$y = \alpha x - (1-\alpha) z \in D_\alpha(x).$$

Similarly we prove $x \in D_{\alpha'}(y)$ for an $\alpha' \geqq 1$. Hence we have proved $x \sim y$, and we are through. □

*If K is a compact convex subset of $\mathbb{R}^n$ and F is a closed convex subset of K, then $A(F;K)$ is finite dimensional. Hence the two norms of $A(F;K)$ are topologically equivalent, and the characteristic number ρ_F is *finite* (cf. (5.11)). In particular, *every Archimedean face of a finite dimensional compact convex set is strongly Archimedean* (Th. II.5.14). However, we have the following:

Proposition II.5.26. *In $\mathbb{R}^{\aleph_0}$ there exists a compact convex set K with a face F which is Archimedean, but not strongly Archimedean.*

Proof. Let E be a countable product of Euclidean planes, i.e $E = (\mathbb{R}^2)^N$ with product topology, and let $K = \prod_{n=1}^{\infty} K_n$, $F = \prod_{n=1}^{\infty} F_n$, where

$$K_n = \left\{ (\xi, \eta) \mid 0 \leqq \xi \leqq 1,\ 0 \leqq \eta \leqq \frac{1}{n}((n-1)\,\xi + 1) \right\}, \tag{5.30}$$

$$F_n = \left\{ (0, \eta) \mid 0 \leqq \eta \leqq \frac{1}{n} \right\}. \tag{5.31}$$

Geometrically, K is a product of trapezoids K_n which become more and more oblique as n increases: while F is the product of the shorter edges F_n of these trapezoids.

We denote the elements of E by $x = \{(\xi_i, \eta_i)\}_{i=1,2,\ldots}$, and we claim that $A(K)$ consists of all functions a of the form

$$a(x) = \alpha_0 + \sum_{i=1}^{\infty} (\alpha_i \xi_i + \beta_i \eta_i), \tag{5.32}$$

where

$$\sum_{i=1}^{\infty} (|\alpha_i| + |\beta_i|) < \infty. \tag{5.33}$$

Clearly, every $a \in A(K)$ is of the form (5.32) for some sequence $\{(\alpha_i, \beta_i)\}_{i=1,2,\ldots}$ of coefficients. To verify (5.33) we evaluate a at the points $x, y \in K$, defined by:

$$x=\{(\psi(\alpha_i), \psi(\beta_i)\}_{i=1,2,\ldots}; \qquad y=\{(1-\psi(\alpha_1), 1-\psi(\beta_i))\}_{i=1,2,\ldots},$$

where $\psi(\xi)=1$ if $\xi \geqq 0$, and $\psi(\xi)=0$ if $\xi<0$.

Now:

$$a(x)=\alpha_0+\sum_{i=1}^{\infty}(\alpha_i^+ + \beta_i^+)$$

while

$$a(y)=\alpha_0+\sum_{i=1}^{\infty}(\alpha_i^- + \beta_i^-).$$

Hence

$$\sum_{i=1}^{\infty}(|\alpha_i|+|\beta_i|)=a(x)-a(y)<\infty .$$

The proof that every function a of the form (5.32), (5.33) is continuous is straightforward.

To prove that F is Archimedean, we assume that $a \in A(K)$ and that $a|F \geqq 0$. Let a be represented in the form (5.32), (5.33), and define a new function c on K by

$$c(x)=\alpha_0+\sum_{i=1}^{\infty}[(\alpha_i^+ \vee \beta_i^-)\xi_i+\beta_i\eta_i]$$

Clearly, c is of the same form (5.32), (5.33), and so $c \in A(K)$. Clearly also $c|F=a|F$. We claim that $c \geqq a$ and $c \geqq 0$. To prove this claim, we first define $J=\{i \mid \beta_i \geqq 0\}$ and evaluate the function a at the point $\{(0, \eta_i)\}_{i=1,2,\ldots}$, where $\eta_i=0$ if $i \in J$ and $\eta_i=1/i$ if $i \notin J$. Now the condition $a|F \geqq 0$ yields

$$\alpha_0-\sum_{i \notin J}\frac{\beta_i}{i} \geqq 0 .$$

We denote the left-hand term of this inequality by γ_0, and decompose the function a as follows:

$$a(x)=\gamma_0+\sum_{i=1}^{\infty}a_i(\xi_i, \eta_i),$$

where

$$a_i(\xi_i, \eta_i)=\begin{cases}\alpha_i\xi_i+\beta_i\eta_i & \text{if } i \in J,\\ \alpha_i\xi_i+\beta_i\left(\dfrac{1}{i}-\eta_i\right), & \text{if } i \notin J.\end{cases}$$

Similarly we may decompose c as follows

$$c(x)=\gamma_0+\sum_{i=1}^{\infty} c_i(\xi_i,\eta_i),$$

where

$$c_i(\xi_i,\eta_i)=\begin{cases}(\alpha_i^+\vee\beta_i^-)\,\xi_i+\beta_i\,\eta_i, & \text{if } i\in J,\\ (\alpha_i^+\vee\beta_i^-)\,\xi_i+\beta_i^-\left(\dfrac{1}{i}-\eta_i\right), & \text{if } i\notin J.\end{cases}$$

By evaluating a_i and c_i at the vertices (1,0), (1,1) of the trapezoid K_i, one can verify that $c_i\geqq a_i$ and $c_i\geqq 0$ for all $i\in N$. (The cases $i\in J$ and $i\notin J$ should be treated separately.) It follows that $c\geqq a$, $c\geqq 0$, and the Archimedicity is proved.

Finally we consider the functions $d_n\in A(K)$ defined by

$$d_n(x)=n\eta_n,\qquad n=1,2,\dots,$$

where $x=\{(\xi_i,\eta_i)\}_{i=1,2,\dots}$, as before.

Clearly $\|d_n\|_F=1$ for all n, while $\|d_n\|_q\to\infty$ as $n\to\infty$. It follows that $\rho_F=+\infty$ (cf. (5.11)). Hence F is not strongly Archimedean. ▯

Notes. The material of this section is the work of many hands, and it is difficult to specify the origin of the various notions.—Order ideals were used by Kadison in his 1951 memoir, where he established a connection between maximal order ideals and extreme points and applied this in a unified treatment of functional representation of various partially ordered topological algebraic structures [224].—The investigation of order ideals was continued by Bonsall in a series of papers from the mid-fifties [77], [78], [79], [80]. Among these we should like to point out the 1956 paper on "Extremal maximal ideals", where the perfect maximal ideals are introduced and studied [79].—There is a classical result on partially ordered normed spaces, which was found by Grosberg and Krein in 1939 [198]. They showed that there is a close connection between the supremum value of the norm over order intervals and the best possible bound λ in decompositions of the type (5.7) in the dual space. Our Lemma II.5.4 is merely a special case of the general theorem of Grosberg and Krein.—Partially ordered normed spaces have been studied more recently by many authors, but most of this research is not directly connected with the subject matter of this book. However, we should like to mention Ellis' 1964 paper on "The duality of partially ordered normed linear spaces" [164]. Here he studies problems related to the Krein-Grosberg result, and he also proves that if an order-unit space is a Banach dual space (i.e. if the unit ball is compact in some locally convex Hausdorff topology), then it admits a base-normed pre-dual. The proof of this fact is of some interest in itself, invoking the Tukey-Klee-Ellis' Lemma that the convex hull C of a finite number of bounded closed convex subsets of a Banach space contains every open sphere in which it is dense [388], [244], [164].—Theorem II.5.5 was established by Edwards in 1964 [142].—The facial structure of a simplex was studied to some extent by Alfsen in his 1964 paper "On the geometry of simplexes". [5]. He introduced the concept of a self-determining face, and proved that a closed face of a Bauer simplex is self-determining. However, the problem for general simplexes was left open, and it was settled some two years later by Edwards and Lazar, independently [144], [258], [261].—Theorem II.5.9 was proved by Alfsen in 1968 in a slightly less general form [12].—The concepts of an "Archimedean ideal" and of an "Archimedean face" were defined by Størmer in 1968 [382]. He also proved Theorem II.5.13 under slightly more restrictive hypotheses.—The concepts of "strong Archimedicity" and "characteristic number" are due to Alfsen, and so are Theorem II.5.14 and Proposition II.5.26. [12].—The concept of a "part" was first introduced for function algebras by Gleason in 1957 [184], and the connection with convexity has been clearified in a series of papers by Bear from the years 1965–68 [50], [51], [52], [55]. Theorem II.5.24 was proved in 1967 by Bear and Weiss [57].—

§ 6. Split-faces and Facial Topology

In the present chapter we shall continue the investigation of the facial structure of a compact convex set K, and again we shall assume that K is regularly embedded in a locally convex Hausdorff space E. Hence $E \cong A(K)^*$, and K is located on a hyperplane H which does not pass through the origin of E (cf. Ch. II. § 2).

We shall say that two convex subsets B, D of K are *affinely independent* if every point x of their convex hull can be expressed by a unique convex combination:

$$x = \lambda y + (1-\lambda) z; \quad 0 \leqq \lambda \leqq 1, \quad y \in B, \quad z \in D. \tag{6.1}$$

In this case we shall say that the set $G = \mathrm{conv}(B \cup D)$ is *direct convex sum* of B and D, and we shall write

$$G = B \oplus_c D. \tag{6.2}$$

Proposition II.6.1. *Two convex subsets B,D of K are affinely independent iff their linear hulls satisfy:*

$$\operatorname{lin} B \cap \operatorname{lin} D = \{0\}. \tag{6.3}$$

Proof. 1. Assume first that B,D are affinely independent. An arbitrary point $z \in \operatorname{lin} B \cap \operatorname{lin} D$ can be expressed by

$$z = \alpha_1 u_1 - \alpha_2 u_2 = \beta_1 v_1 - \beta_2 v_2, \tag{6.4}$$

where $\alpha_i, \beta_i \geqq 0$, $u_i \in B$, $v_i \in D$ for $i=1,2$.

For some real number γ we shall have $z \in \gamma H$. Then

$$\gamma = \alpha_1 - \alpha_2 = \beta_1 - \beta_2,$$

and so we may define a positive number δ by writing

$$\delta = \alpha_1 + \beta_2 = \alpha_2 + \beta_1. \tag{6.5}$$

If $\delta = 0$, then $\alpha_i = \beta_i = 0$ for $i=1,2$. Hence $z=0$ as desired.

If $\delta > 0$, then we may rewrite (6.5) as follows

$$\alpha_1 \delta^{-1} u_1 + \beta_2 \delta^{-1} v_2 = \alpha_2 \delta^{-1} u_2 + \beta_1 \delta^{-1} v_1. \tag{6.6}$$

Since B and D are affinely independent, these two convex combinations must be identical. It follows that $\alpha_1 u_1 = \alpha_2 u_2$ and $\beta_2 v_2 = \beta_1 v_1$. Hence $z=0$ also in this case.

2. Assume next that $\operatorname{lin} B \cap \operatorname{lin} D = \{0\}$, and consider an equation

$$\lambda y + (1-\lambda) z = \lambda' y' + (1-\lambda') z', \tag{6.7}$$

where $0 \leqq \lambda, \lambda' \leqq 1$; $y, y' \in B$; $z, z' \in D$.

By hypothesis

$$\lambda y - \lambda' y' = (1-\lambda') z' - (1-\lambda) z = 0.$$

Since $y, y' \in H$, the equation $\lambda y = \lambda' y'$ can only be satisfied if $\lambda = \lambda'$, and $y = y'$ when $\lambda \neq 0$, and $z = z'$ when $\lambda \neq 1$. Hence the two convex combinations of (6.7) are identical.

We recall that an *affine dependence* on a compact convex set K is a (signed) measure μ such that $\mu(a)=0$ for all $a \in A(K)$ (cf. Ch. I, § 2). The linear space of affine dependences on K is denoted by either of the two symbols $N(K)$ or $A(K)^{\perp}$. Recall also that a boundary measure in $A(K)^{\perp}$ is said to be a *boundary affine dependence* (cf. Ch. II, § 4). For convenience we shall use the symbol μ_B to denote the "restriction" of a measure μ on K to a Borel set $B \subset K$. Thus $\mu_B(C) = \mu(B \cap C)$ for every Borel set $C \subset K$.

For Borel subsets of K it is sometimes convenient to use a notion of convexity which is stronger than ordinary convexity. We shall say that a Borel set $B\subset K$ is *measure convex* if

$$\mu\in M_1^+(K), \quad \mu(B)=1 \Rightarrow \text{barycenter}(\mu)\in B. \tag{6.8}$$

Clearly every measure convex Borel set is convex. (It is even *σ-convex* in the sense defined at the end of Ch. I, § 4). Note also that *every closed convex set is measure convex* (Prop I.2.1). But it is *not* so that every convex Borel set is measure convex. Consider for example $K=M_1^+([0,1])$ and let B be the convex set of all purely atomic measures in K. Now B is a convex Borel set containing $\overline{\partial_e K}$ (cf. Ex. I.2.10, and recall that $\partial_e K=\{\varepsilon_t \mid t\in[0,1]\}$). It follows that B is not measure convex, since a measure convex Borel set containing $\overline{\partial_e K}$ must be all of K.

Proposition II.6.2. *Let B and D be two disjoint measure convex Borel subsets of K. Then B and D are affinely independent iff.*

$$\mu\in A(K)^\perp, \quad |\mu|(K\backslash(B\cup D))=0 \Rightarrow \mu_B,\mu_D\in A(K)^\perp. \tag{6.9}$$

Proof. 1. Assume first that B,D are affinely independent, and consider a measure $\mu\in A(K)^\perp$ such that $|\mu|(K\backslash(B\cup D))=0$. By Jordan decomposition we can write $\mu=\mu_1-\mu_2$ where $\mu_1,\mu_2\geqq 0$, and without lack of generality we can assume that μ_1,μ_2 are probability measures.

We write $\gamma_i=\mu_i(B)$ and we first assume that $0<\gamma_i<1$ for $i=1,2$. Now we define y_i to be the barycenter of the probability measure $\gamma_i^{-1}(\mu_i)_B$ and z_i to be the barycenter of the probability measure $(1-\gamma_i)^{-1}(\mu_i)_D$ for $i=1,2$.

By the assumption $\mu\in A(K)^\perp$ and by the definition of barycenter, we obtain for every $a\in A(K)$:

$$0=\mu_1(a)-\mu_2(a)=\gamma_1 a(y_1)+(1-\gamma_1)a(z_1)-\gamma_2 a(y_2)-(1-\gamma_2)a(z_2).$$

Since $a\in A(K)$ was arbitrary, this implies:

$$\gamma_1 y_1+(1-\gamma_1)z_1=\gamma_2 y_2+(1-\gamma_2)z_2. \tag{6.10}$$

By measure convexity, $y_i\in B$ and $z_i\in D$ for $i=1,2$. Now the affine independence of B and D entails $y_1=y_2$ and $z_1=z_2$. Hence for every $a\in A(K)$

$$0=a(y_1)-a(y_2)=\mu_B(a); \qquad 0=a(z_1)-a(z_2)=\mu_D(a),$$

and so we have proved $\mu_B,\mu_D\in A(K)^\perp$ in this case.

It remains to study the exceptional cases $\gamma_i=0,1$ for $i=1,2$, and by symmetry it suffices to study the case $\gamma_1=0$. We claim that also

$\gamma_2=0$, which will complete the proof, giving $\mu_B=0\in A(K)^\perp$, $\mu_D=\mu\in A(K)^\perp$.

To prove this claim we first observe that $\mu_2(D)>0$, for otherwise $\mu_2(B)=1$ and the common barycenter of μ_1 and μ_2 would be in $B\cap D$, which is impossible since $B\cap D=\emptyset$. Hence $0\leqq\gamma_2<1$. We denote the barycenter of μ_1 by z_1 and assume for contradiction $0<\gamma_2<1$. Then we can define y_2, z_2 as before, and we obtain the following "degenerate version of formula (6.10)":

$$z_1=\gamma_2 y_2+(1-\gamma_2)z_2. \tag{6.11}$$

Here $z_1\in D$, $y_2\in B$, $z_2\in D$, and the affine independence of B and D entails $\gamma_2=0$ which is the desired contradiction.

2. Assume next that (6.9) is valid, and consider an equation

$$\lambda y+(1-\lambda)z=\lambda' y'+(1-\lambda')z', \tag{6.12}$$

where $0\leqq\lambda,\lambda'\leqq 1$; $y,y'\in B$; $z,z'\in D$.

The measure

$$\nu=(\lambda\varepsilon_y-\lambda'\varepsilon_{y'})+((1-\lambda)\varepsilon_z-(1-\lambda')\varepsilon_{z'})$$

is an affine dependence such that $|\nu|(K\backslash(B\cup D))=0$. By assumption $\nu_B,\nu_D\in A(K)^\perp$. A two-point measure can never be an affine dependence (since $A(K)$ separates points). Hence we shall have $\lambda=\lambda'$, and $y=y'$ if $\lambda\neq 0$, and $z=z'$ if $\lambda\neq 1$. Thus the two convex combinations of (6.12) are identical, and the proof is complete. □

Our next proposition is somewhat technical, pertaining to a rather special situation in which the values of the upper envelope $\hat{f}$ of a given function f on K can be effectively obtained by (finite) convex combinations of values of f. (Cf. Cor. I.3.6.) In this connection we shall need the following:

Lemma II.6.3. *A function $f\colon K\to\mathbb{R}^+$ is u.s.c. iff its positive subgraph*

$$\mathrm{Sub}(f)=\{(y,\alpha)\,|\,y\in K,\ 0\leqq\alpha\leqq f(y)\} \tag{6.13}$$

is a closed subset of $K\times\mathbb{R}^+$; f is a concave function iff Sub(f) *is a convex set, and generally*

$$\mathrm{Sub}(\hat{f})=\mathrm{cl.conv}[\mathrm{Sub}(f)]. \tag{6.14}$$

The proof is an elementary verification (for details cf. e.g. Prop. 2.1–2.2 and Th. 3.9 of [91]). □

Proposition II.6.4. *Let F_j be closed faces of K and let b_j be real valued functions on K such that $b_j|F_j \in A(F_j)^+$, $b_j|(K\backslash F_j)=0$ for $j=1,\dots,m$. Also let $a_i \in A(K)^+$ for $i=1,\dots,n$. For every $x \in K$ there exists a convex combination*

$$x = \sum_{i=1}^{n} \lambda_i x_i + \sum_{j=1}^{m} \mu_j y_j, \tag{6.15}$$

such that $y_j \in F_j$ for $j=1,\dots,m$, and

$$\overline{a_1 \vee \cdots \vee a_n \vee b_1 \vee \cdots \vee b_n}(x) = \sum_{i=1}^{n} \lambda_i a_i(x_i) + \sum_{j=1}^{m} \mu_j b_j(y_j). \tag{6.16}$$

Proof. We define subsets of $K \times \mathbb{R}$ as follows

$$Q_i = \{(y,\alpha) | y \in K,\ 0 \leqq \alpha \leqq a_i(y)\}, \quad i=1,\dots,n,$$
$$R_j = \{(y,\alpha) | y \in F_j,\ 0 \leqq \alpha \leqq b_j(y)\}, \quad j=1,\dots,m.$$

Clearly Q_i, R_j are compact convex sets and

$$\mathrm{Sub}(a_1 \vee \cdots \vee a_n \vee b_1 \vee \cdots \vee b_m) = \bigcup_{i=1}^{n} Q_i \cup \bigcup_{j=1}^{m} R_i.$$

Recalling that the convex hull of a finite union of compact convex sets is compact, we can apply Lemma II.6.5 to obtain:

$$\overline{\mathrm{Sub}(a_1 \vee \cdots \vee a_n \vee b_1 \vee \cdots \vee b_m)} = \mathrm{conv}\left(\bigcup_{i=1}^{n} Q_i \cup \bigcup_{j=1}^{m} R_j \right).$$

From this the proposition follows. □

To every subset G of K we associate its *complementary set* G' which is the union of all faces disjoint from G. In symbols:

$$x \in G' \Leftrightarrow \mathrm{face}(x) \cap G' = \emptyset.$$

The set G' is not convex in general; but if it is convex, then it is seen to be a face.

The complementary set F' of a *closed* face F can be determined by means of the upper envelope of the characteristic function χ_F:

Proposition II.6.5. *If F is a closed face of K, then for every $x \in K$ there exists a convex combination*

$$x = \lambda y + (1-\lambda) z, \quad \text{where } y \in F, z \in F', \lambda = \hat{\chi}_F(x). \tag{6.17}$$

From this it follows that

$$\hat{\chi}_F^{-1}(1)=F, \quad \hat{\chi}_F^{-1}(0)=F', \quad K=\mathrm{conv}(F\cup F').$$

In particular F' is a G_δ-set.

Proof. 1. By Proposition II.6.4, $\hat{\chi}_F(x)=\lambda$ for some convex combination

$$x=\lambda y+(1-\lambda)z, \quad \text{where } y\in F. \tag{6.18}$$

If $\lambda=1$, then z is arbitrary, and we can as well choose $z\in F'$, obtaining a convex combination with all the desired properties.

Henceforth we assume $\lambda<1$, and we shall prove that the point z of (6.18) must necessarily be in F'.

Observe first that $\hat{\chi}_F(z)=0$, since an application of the concave function $\hat{\chi}_F$ to (6.18) gives

$$\lambda\geqq\lambda+(1-\lambda)\hat{\chi}_F(z).$$

We assume for contradiction that $z\notin F'$, say $u\in\mathrm{face}(z)\cap F$. Then there must exist a convex combination

$$z=\gamma u+(1-\gamma)v, \quad \text{where } \gamma\neq 0,$$

and an application of the concave function $\hat{\chi}_F$ gives the desired contradiction

$$0\geqq\gamma+(1-\gamma)\hat{\chi}_F(v)\geqq\gamma>0.$$

2. If $x\in F$, then trivially $\hat{\chi}_F(x)=1$. If $x\notin F$, then we must have

$$\hat{\chi}_F(x)=\lambda<1.$$

3. If $x\in F'$, then we must have $\lambda=0$ in (6.17), and so $\hat{\chi}_F(x)=0$. If $x\notin F'$, then we must have $\lambda>0$ in (6.17), and so $\hat{\chi}_F(x)>0$.

4. It is clear from (6.17) that $K=\mathrm{conv}(F\cup F')$.

5. Finally F' is a G_δ-set since $\hat{\chi}_F$ is u.s.c. □

A proper face F of K is said to be a *split face* if F' is a face (i.e. if it is convex) and if K is direct convex sum of F and F'. The empty set $\emptyset$ and the whole set K are said to be *(improper) split faces*. If F is a split face, then F' is also a split face, and we shall say that (F,F') is a pair of complementary split faces.

Proposition II. 6.6. *If F,G are proper split faces of K, then every $x\in K$ can be decomposed as*

$$x=\alpha_{11}x_{11}+\alpha_{12}x_{12}+\alpha_{21}x_{21}+\alpha_{22}x_{22}, \tag{6.19}$$

where $\alpha_{ij} \geqq 0$ *for* $i=1,2$, $\sum_{i,j=1}^{2} \alpha_{ij}=1$ *and* $x_{11} \in F \cap G$, $\chi_{12} \in F \cap G'$, $x_{21} \in F' \cap G$, $x_{22} \in F' \cap G'$. *This decomposition is unique, in that every* α_{ij} *is uniquely determined and every* x_{ij} *with non-vanishing coefficient* α_{ij} *is uniquely determined.*

Proof. We obtain a decomposition of the type (6.19) by decomposing first with respect to F, F' and then the obtained F, F'-components with respect to G, G'.

To prove that the decomposition (6.19) is unique, we write $y_{ij}=\alpha_{ij} x_{ij}$ for $i,j=1,2$. Now $x=\sum_{i\,j=1}^{2} y_{ij}$, and $(y_{11}+y_{12}) \in \operatorname{lin} F$, $(y_{21}+y_{22}) \in \operatorname{lin} F'$. Since F and F' are affinely independent, we shall have $\operatorname{lin} F \cap \operatorname{lin} F'=\{0\}$ (Prop. II.6.1). It follows that each of the two vectors $y_{11}+y_{12}$ and $y_{21}+y_{22}$ is uniquely determined. Repeating the same argument with G, G' in the place of F, F', we conclude that each of the four vectors $y_{11}, y_{12}, y_{21}, y_{22}$ is uniquely determined.

Finally we recall that $y_{ij}=\alpha_{ij} x_{ij}$, and $x_{ij} \in H$ for i,j. Hence every α_{ij} is uniquely determined, and x_{ij} is uniquely determined when $\alpha_{ij} \neq 0$. This completes the proof. □

Proposition II.6.7. *If F and G are split faces of K, then* $F \cap G$ *is also a split face, and*

$$(F \cap G)'=\operatorname{conv}(F' \cup G'). \tag{6.20}$$

Proof. The statement is trivial if F or G, or both, are improper. Hence we may assume that F, G are proper split faces. By Proposition II.6.6 it suffices to prove that the set complementary to $F \cap G$ is the convex hull of $F' \cup G'$.

To this end we first assume that $x \in \operatorname{conv}(F' \cup G')$, and we assume for contradiction that there exists a convex combination

$$x=\alpha y+(1-\alpha) z, \qquad y \in F \cap G,\ \alpha \neq 0. \tag{6.21}$$

Decomposing z as in (6.19), we obtain

$$x=(\alpha y+(1-\alpha) \beta_{11} z_{11})+(1-\alpha)(\beta_{12} z_{12}+\beta_{21} z_{21}+\beta_{22} z_{22}).$$

Writing $\delta=\alpha+(1-\alpha) \beta_{11}$ we shall have $\delta \neq 0$, since $\alpha \neq 0$. Now

$$u=\delta^{-1}(\alpha y+(1-\alpha) \beta_{11} z_{11}) \in F \cap G,$$

and the decomposition

$$x=\delta u+(1-\alpha)(\beta_{12} z_{12}+\beta_{21} z_{21}+\beta_{22} z_{22})$$

is of the type (6.19). Now it follows from the assumption $x \in \operatorname{conv}(F' \cup G')$ that $\delta = 0$, which is a contradiction. Hence there can not be any convex combination of the type (6.21), and so $x \in (F \cap G)'$.

Next we assume that $x \notin \operatorname{conv}(F' \cup G')$. Then the coefficient α_{11} in the decomposition (6.19) of x must be non-zero. Hence $x_{11} \in \operatorname{face}(x) \cap F \cap G$, and so $x \notin (F \cap G)'$. This completes the proof. □

Corollary II.6.8. *If F and G are non-empty split-faces of K, then* $\operatorname{conv}(F \cup G)$ *is also a split-face.*

We now pass to the study of *closed* split faces and their complements.

Proposition II.6.9. *If F is a closed split face of K, then $\hat{\chi}_F$ is affine.*

Proof. By Proposition II.6.5, $\hat{\chi}_F(x) = \lambda$ where $x \in K$ is an arbitrary point of K and λ is the coefficient occuring in the unique convex combination

$$x = \lambda y + (1 - \lambda) z, \quad y \in F, \quad z \in F'.$$

From this the proposition follows. □

Remark. The above property does not characterize split faces. If F is an edge of a square, then $\hat{\chi}_F$ is affine, but F is not a split face.

Lemma II.6.10. *Let F be a face of K which is either closed or of the form $F = f^{-1}(0)$ where f is an u.s.c. positive affine function; and let μ be a probability measure on K with barycenter x. Then*

$$\mu(F) = 1 \Leftrightarrow x \in F. \tag{6.22}$$

Proof. 1. Assume first that F is closed. If $\mu(F) = 1$, then $x \in F$ by Proposition I.2.1.

If $x \in F$, then we approximate μ (vaguely) by simple measures with barycenter x (Prop. I.2.3). They are all supported by F since F is a face, and the limiting measure μ is also supported by F since F is closed.

2. Assume next that $F = f^{-1}(0)$ where f is u.s.c., positive and affine. It is easily verified that the barycenter formula holds for f. (Apply Cor. I.1.4 and a limiting argument based on formula (2.3) of Ch. I.). Thus, $\mu(f) = f(x)$, and the equation (6.22) follows. □

Corollary II.6.11. *If F is a closed split face of K, then the equivalence (6.22) is valid, and it is also valid with F' in the place of F. In particular, F' is measure convex.*

Proof. The function $\hat{\chi}_F$ is u.s.c., affine (Prop. II.6.9), and $F'=\hat{\chi}_F^{-1}(0)$ (Prop. II.6.5).

Hence we may apply the proposition above. □

Theorem II.6.12. *If F is a closed face of K, then the following statements are equivalent.*

(i) *F is a split face.*

(ii) $\mu_F \in A(K)^\perp$ *for every affine dependence* μ *concentrated on* $F \cup F'$.

(iii) $\mu_F \in A(K)^\perp$ *for every boundary affine dependence* μ *on* K.

Proof. (i) ⇒ (ii) We first assume that F is a split face and consider a measure $\mu \in A(K)^\perp$ such that $|\mu|(K \backslash (F \cup F'))=0$. The sets F and F' are measure convex and affinely independent. Hence we can apply Proposition II.3.6 to obtain $\mu_F \in A(K)^\perp$.

(ii) ⇒ (iii) We next assume (ii) and consider a boundary measure $\mu \in A(K)^\perp$. By the "Remark" to Proposition I.4.5, applied to the u.s.c. function χ_F, we shall have $|\mu|(\hat{\chi}_F)=|\mu|(\chi_F)$. Hence by Proposition II.6.5:

$$|\mu|(K \backslash (F \cup F'))=|\mu|(\{x \mid \hat{\chi}_F(x)>\chi_F(x)\})=0.$$

Now we can apply (ii) to obtain $\mu_F \in A(K)^\perp$.

(iii) ⇒ (i) By condition (iii) and Lemma II.6.10 $x \in F'$ iff $\mu(F)=0$ for one, and hence for all, boundary measures $\mu \in M_x^+(K)$. From this the convexity of F' follows.

Finally we consider an equation

$$\lambda_1 y_1+(1-\lambda_1) z_1=\lambda_2 y_2+(1-\lambda_2) z_2, \tag{6.23}$$

where $0 \leqq \lambda_i \leqq 1$, $y_i \in F$, $z_i \in F'$ for $i=1,2$. Without lack of generality we also assume $\lambda_1>0$.

For $i=1,2$ we choose boundary measures $\mu_i, \nu_i \in M_1^+(K)$ representing y_i, z_i, respectively. By formula (6.22) we shall have $\mu_i(F)=1, \nu_i(F')=1$, $i=1,2$. The measure

$$\pi=(\lambda_1 \mu_1-\lambda_2 \mu_2)+((1-\lambda_1) \nu_1-(1-\lambda_2) \nu_2)$$

is seen to be an affine dependence, and by the assumption (iii) we shall have

$$\pi_F=(\lambda_1 \mu_1-\lambda_2 \mu_2) \in A(K)^\perp. \tag{6.24}$$

Evaluating at $1_K \in A(K)$, we obtain $\lambda_2=\lambda_1>0$. Hence $\mu_1-\mu_2 \in A(K)^\perp$, and it follows that μ_1 and μ_2 have common barycenter. Thus, the two convex combinations of (6.23) are identical, and the affine independence of F and F' is proved. □

We proceed to show that the continuous affine functions on a closed split face enjoy a strong extension property, and we start by some lemmas:

Lemma II.6.13. *Let F be a closed split face of K and let b be a real valued function on K such that $b|F\in A(F)^+$ and $b|(K\setminus F)=0$. Also let $a_1,\dots,a_n\in A(K)^+$, and define $f=b\vee a_1\vee\cdots\vee a_n$. If x is an arbitrary point in K with decomposition*

$$x=\lambda y+(1-\lambda)z;\quad 0\leqq\lambda\leqq 1,\quad y\in F,\quad z\in F', \tag{6.25}$$

then

$$\hat{f}(x)=\lambda\hat{f}(y)+(1-\lambda)\hat{f}(z). \tag{6.26}$$

Proof. 1. Trivially $\hat{f}(x)\geqq\lambda\hat{f}(y)+(1-\lambda)\hat{f}(z)$.

2. By Proposition II.6.4, there is a convex combination

$$x=\sum_{i=0}^{n}\lambda_i x_i;\quad x_0\in F,\quad x_1,\dots,x_n\in K, \tag{6.27}$$

such that

$$\hat{f}(x)=\lambda_0 b(x_0)+\sum_{i=1}^{n}\lambda_i a_i(x_i). \tag{6.28}$$

Now we decompose each x_i with resp. to F and F':

$$x_i=\alpha_i y_i+(1-\alpha_i)z_i, \tag{6.29}$$

where $y_i\in F$, $z_i\in F'$, and $0\leqq\alpha_i\leqq 1$ for $i=0,\dots,n$. In particular $\alpha_0=1$ and $y_0=x_0$ since $x_0\in F$.

Define $\alpha=\sum_{i=0}^{n}\lambda_i\alpha_i$, and assume first $0<\alpha<1$.

Now by (6.27):

$$x=\alpha\left[\sum_{i=0}^{n}\alpha^{-1}\lambda_i\alpha_i y_i\right]+(1-\alpha)\left[\sum_{i=1}^{n}(1-\alpha)^{-1}\lambda_i(1-\alpha_i)z_i\right],$$

and this is a decomposition of x with resp. to F and F'. By uniqueness we get $\alpha=\lambda$ and

$$y=\lambda^{-1}\sum_{i=0}^{n}\lambda_i\alpha_i y_i,\qquad z=(1-\lambda)^{-1}\sum_{i=1}^{n}\lambda_i(1-\alpha_i)z_i.$$

These expressions are both convex combinations (recall that $\lambda=\alpha=\sum_{i=0}^{n}\lambda_i\alpha_i$), and the concave function $\hat{f}$ will satisfy the inequalities:

$$\hat{f}(y)\geqq\lambda^{-1}\left(\lambda_0 b(x_0)+\sum_{i=1}^{n}\lambda_i\alpha_i a_i(y_i)\right), \tag{6.30}$$

$$\hat{f}(z)\geqq(1-\lambda)^{-1}\sum_{i=1}^{n}\lambda_i(1-\alpha_i)a_i(z_i). \tag{6.31}$$

By application of (6.28), (6.29), and then of (6.30), (6.31), we obtain

$$\hat{f}(x)=\lambda_0 b(x_0)+\sum_{i=1}^{n}\lambda_i\alpha_i a_i(y_i)+\sum_{i=1}^{n}\lambda_i(1-\alpha_i)^{-1}a_i(z_i)\leqq\lambda\hat{f}(y)+(1-\lambda)\hat{f}(z).$$

This completes the proof if $0<\alpha<1$.

If $\alpha=0,1$, then $x\in F$ or $x\in F'$, and there is nothing to prove. □

Lemma II.6.14. *Let F be a closed split face of K, let $a_1,\dots,a_n,b_1\in A(K)$, let $a_0,b_0\in A(F)$, and assume that*

$$a_i<b_1,\qquad a_i|F\leqq a_0<b_0\leqq b_1|F \tag{6.32}$$

for $i=1,\dots,n$. Then there is a $c\in A(K)$ such that

$$a_i<c<b_1,\qquad a_0<c|F<b_0 \tag{6.33}$$

for $i=1,\dots,n$.

Proof. Let $[\alpha,\beta]$ be some compact interval containing the range of all occuring functions $a_0,\dots,a_n$, b_0,b_1. Define a_0' as a_0 on F and α on $K\backslash F$, and define b_0' as b_0 on F and β on $K\backslash F$. Let

$$a=a_0'\vee a_1\vee\cdots\vee a_n,\qquad b=b_0'\wedge b_1\,. \tag{6.34}$$

We claim that

$$\hat{a}<\hat{b}\,. \tag{6.35}$$

Observe first that $\hat{a}(x)<\check{b}(x)$ if $x\in F$ or $x\in F'$. This follows by Proposition II.6.4, since the envelopes are obtained by finite convex combinations, and F,F' are faces.

Next we consider a point $x\in K\backslash(F\cup F')$ with a decomposition

$$x=\lambda y+(1-\lambda)(z),\qquad y\in F,\qquad z\in F',\qquad 0<\lambda<1\,.$$

By Lemma II.6.13 and by the validity of (6.35) on F and F':

$$\hat{a}(x)=\lambda\hat{a}(y)+(1-\lambda)\hat{a}(z)<\lambda\check{b}(y)+(1-\lambda)\check{b}(z)=\check{b}(x)\,. \tag{6.36}$$

By an easy application of Hahn-Banach separation in product space (cf. Prop. I.1.2), it follows that there exists a $c\in A(K)$ such that $\hat{a}<c<\check{b}$. This function c will have all the desired properties, and the proof is complete. □

Theorem II.6.15. *Let F be a closed split face of K, let $a_1,\dots,a_n$, $b\in A(K)$, let $a_0\in A(F)$, and assume that*

$$a_i\leqq b,\qquad a_i|F\leqq a_0\leqq b|F\,, \tag{6.37}$$

for $i=1,\ldots,n$. *Then for every* $\varepsilon>0$ *there is a* $c\in A(K)$ *such that*

$$a_i \leqq c \leqq b+\varepsilon, \qquad c|F=a_0 \tag{6.38}$$

for $n=1,\ldots,n$.

Proof. We shall construct a Cauchy sequence $\{c_n\}$ from $A(K)$ which converges to a function $c\in A(K)$ satisfying (6.38).

Writing $a=a_1\vee\cdots\vee a_n$ we have

$$a-\varepsilon<b,$$

$$(a-\varepsilon)\,|\,F\leqq a_0-\varepsilon<a_0-\frac{\varepsilon}{2}\leqq b\,|\,F\,.$$

Now we may apply Lemma II.6.14 with $a_i-\varepsilon$ in the place of a_i for $i=1,\ldots,n$, with b in the place of b_1, with $a_0-\varepsilon$ in the place of a_0, and with $a_0-\varepsilon/2$ in the place of b_0. Hence there is a function $c_1\in A(K)$ such that

$$a-\varepsilon<c_1<b. \tag{6.39}$$

$$a_0-\varepsilon<c_1\,|\,F<a_0-\frac{\varepsilon}{2}. \tag{6.40}$$

By continuity and compactness there exists a number $\gamma_1\in\,]0,\varepsilon/4[$ such that

$$a_0-\varepsilon+\gamma_1<c_1|F\,.$$

Now we shall have

$$c_1\vee\left(a-\frac{\varepsilon}{2}\right)<c_1+\frac{\varepsilon}{2},$$

and also

$$c_1|F\vee\left(a-\frac{\varepsilon}{2}\right)\Big|\,F\leqq a_0-\frac{\varepsilon}{2}<a_0-\frac{\varepsilon}{2}+\gamma_1\leqq c_1\,|\,F+\frac{\varepsilon}{2}.$$

Again we may apply Lemma II.6.14 with the obvious specifications to obtain a $c_2\in A(K)$ such that

$$c_1\vee\left(a-\frac{\varepsilon}{2}\right)<c_2<c_1+\frac{\varepsilon}{2}, \tag{6.41}$$

$$a_0-\frac{\varepsilon}{2}<c_2\,|\,F<a_0-\frac{\varepsilon}{4}. \tag{6.42}$$

Continuing by induction we obtain a sequence $\{c_n\}$ from $A(K)$ such that

$$c_n\vee\left(a-\frac{\varepsilon}{2^n}\right)<c_{n+1}<c_n+\frac{\varepsilon}{2^n}, \tag{6.43}$$

$$a_0-\frac{\varepsilon}{2^n}<c_{n+1}\,|\,F<a_0+\frac{\varepsilon}{2^{n+1}}. \tag{6.44}$$

It follows from (6.43) that

$$\|c_{n+1}-c_n\| \leqq \frac{\varepsilon}{2^n}, \qquad n=1,2,\ldots . \tag{6.45}$$

Hence $\{c_n\}$ is a Cauchy sequence with a limit c in $A(K)$. Also it follows from (6.43) that $a \leqq c$. Moreover, by (6.39) and (6.43):

$$c=c_1+\sum_{n=1}^{\infty}(c_{n+1}-c_n) \leqq b+\sum_{n=1}^{\infty}\frac{\varepsilon}{2^n}=b+\varepsilon .$$

Finally, it follows from (6.45) that $c|F=a_0$, and the proof is complete. □

Corollary II.6.16. *A closed split face F of K is strongly Archimedean with characteristic number equal to* 1.

Proof. 1. If $a \in A(K)$ and $a|F \geqq 0$, then we may apply Theorem II.6.15 with $n=2$, $a_1=a$, $a_2=0$, $a_0=a|F$, and an arbitrary constant function $b>a$, to obtain a function $c \in A(K)^+$ such that $c \geqq a$ and $c|F=a|F$. Hence, F has the property (5.16) which defines Archimedean faces.

2. To prove that $\rho_F=1$, we consider a function $a_0 \in A(F)$ such that $-1 \leqq a_0 \leqq 1$. Now we apply Theorem II.6.15 with $n=1$, $a_1=-1$, $b=1$, to obtain a function $c \in A(K)$ such that $-1 \leqq c \leqq 1+\varepsilon$ and $c|F=a_0$, where $\varepsilon>0$ is arbitrary. By the definition of ρ_F, we shall have $\rho_F=1$. □

Corollary II.6.17. *If F is a closed split face of K and* z_1, z_2 *are two distinct points of* F' *then there is an* $a \in A(K)$ *such that* $a|F=0$ *and* $a(z_1) \neq a(z_2)$.

Proof. It follows from Theorem II.5.14 that $\operatorname{lin} F$ is a (w^*)-closed linear subspace of $E(\cong A(K)^*)$. Hence $E/\operatorname{lin} F$ is a locally convex Hausdorff space.

We denote the canonical map of E onto $E/\operatorname{lin} F$ by Φ, and we claim that $\Phi(z_1) \neq \Phi(z_2)$.

To prove this claim we assume the contrary, i.e.

$$z_1-z_2=\lambda_1 y_1-\lambda_2 y_2 , \tag{6.46}$$

for some $\lambda_1, \lambda_2 \geqq 0$, $y_1, y_2 \in F$.

Now $\lambda_1=\lambda_2$ since all points of (6.46) are located on H. Writing $\mu=1+\lambda_1=1+\lambda_2$ and noting that $\mu \neq 0$, we shall have

$$\mu^{-1} z_1+\mu^{-1}\lambda_2 y_2=\mu^{-1} z_2+\mu^{-1}\lambda_1 y_1 .$$

By the uniqueness of decompositions with resp. to F and F' we obtain $z_1 = z_2$, contrary to hypothesis.

Now there is by Hahn-Banach a continuous linear functional ψ on $E/\operatorname{lin} F$ which is non-zero on $\Phi(z_1) - \Phi(z_2)$, and the function $a = (\psi \circ \Phi)|K$ will have the desired properties. □

Theorem II.6.18. *A closed face F of K is split iff $\hat{b}$ is affine for every real valued function b on K such that $b|F \in A(F)^+$ and $b|(K \backslash F) = 0$.*

Proof. 1. Let $\hat{b}$ be a function on K with the properties specified in the theorem. To show b affine, it suffices to prove that the set of all $a \in A(K)$ for which $a > b$, is directed downwards. To this end we consider two functions $a_1, a_2 \in A(K)$ such that $a_1, a_2 > b$. Applying the dual version of Theorem II.6.15 (with all inequality signs reversed), and noting that $a_1, a_2 > b - \varepsilon$ for some $\varepsilon > 0$, we obtain a $c \in A(K)$ such that $b < c < a_1, a_2$.

2. Assume next that $\hat{b}$ is affine for every function b with the properties stated in the theorem. In particular $\hat{\chi}_F$ is affine, hence $F' = \hat{\chi}_F^{-1}(0)$ is a convex set, and thus a *face* of K. Generally $K = \operatorname{conv}(F \cup F')$ (Prop. II.6.5), and it remains only to prove uniqueness of the decomposition with resp. to F and F'.

Let

$$x = \lambda_1 y_1 + (1 - \lambda) z_1 = \lambda_2 y_2 + (1 - \lambda_2) z_2, \tag{6.47}$$

where $y_i \in F$, $z_i \in F'$, $0 \leqq \lambda_i \leqq 1$ for $i = 1, 2$. Since $\hat{\chi}_F$ is affine, we must have $\hat{\chi}_F(x) = \lambda_1 = \lambda_2$.

If $y_1 \neq y_2$, then there is an $a_0 \in A(F)^+$ such that $a_0(y_1) \neq a_0(y_2)$. We define

$$b(x) = \begin{cases} a_0(y) & \text{for } y \in F, \\ 0 & \text{for } y \in K \backslash F. \end{cases}$$

By assumption $\hat{b}$ is affine, and it follows from Proposition II.6.4 that $\hat{b}|F' = 0$. Hence we obtain by application of the function $\hat{b}$ to the equation (6.47):

$$\hat{b}(x) = \lambda_1 a_0(y_1) = \lambda_2 a_0(y_2).$$

This is a contradiction since $\lambda_1 = \lambda_2 \neq 0$ and $a_0(y_1) \neq a_0(y_2)$. Hence we must have $y_1 = y_2$, and the uniqueness is proved. □

We shall also state a simple extension theorem which is independent of the preceding lemmas.

Proposition II.6.19. *Let F be a closed split face of K and assume that F' is closed. If $a \in A(F)$ and $b \in A(F')$, then there exists a unique $c \in A(K)$ such that $c|F = a$, $c|F' = b$. Moreover, if $\alpha \leqq a$, $b \leqq \beta$, then $\alpha \leqq c \leqq \beta$.*

Proof. We define the function c by

$$c(x) = \lambda a(y) + (1-\lambda) b(z),$$

where

$$x = \lambda y + (1-\lambda) z, \qquad 0 \leqq \lambda \leqq 1, \qquad y \in F, \qquad z \in F'.$$

Clearly c is affine and $c|F = a$, $c|F' = b$. Observe that $c = \Psi \circ \Phi^{-1}$, where Φ is a biunique map of $F \times F' \times [0,1]$ onto K defined by

$$\Phi(y, z, \lambda) = \lambda y + (1-\lambda) z,$$

while Ψ is a map of $F \times F' \times [0,1]$ into $\mathbb{R}$ defined by

$$\Psi(y, z, \lambda) = \lambda a(y) + (1-\lambda) b(z).$$

For every closed subset D of $\mathbb{R}$ we shall have

$$c^{-1}(D) = \Phi(\Psi^{-1}(D)),$$

and it follows by the continuity of Φ and Ψ, and by the compactness of K, that $c^{-1}(D)$ is a closed subset of K. Hence we have proved that c is continuous.

The inequalities $\alpha \leqq c \leqq \beta$ are obvious, and the proof is complete. □

We are now in the position to define the "facial topology" of the extreme boundary.

Proposition II.6.20. *The collection of closed split faces of K is closed under finite convex hulls and arbitrary intersections.*

Proof. 1. By Corollary II.6.18, the convex hull of finitely many closed split faces is again a closed split face.

2. By Proposition II.6.7, it suffices to prove that $F = \bigcap_\alpha F_\alpha$ is a split face if $\{F_\alpha\}$ is a downwards directed net of closed split faces. For this purpose it suffices to prove that for every $a_0 \in A(F)^+$ the set

$$\{a \in A(K) \mid a > 0,\ a|F > a_0\} \tag{6.48}$$

is directed downwards. (Cf. Th. II.6.8, and the definition of upper envelope of a function in Ch. I, § 1.)

To prove the directedness of the set described in (6.48), we consider two functions

$$a_i \in A(K),\ a_i > 0,\ a_i|F_i > a_0; \qquad i=1,2.$$

Let g be the real valued function on K which is defined by $g|F = a_0$, $g|(K \backslash F) = 0$. By Proposition II.6.4 we shall have

$$\hat{g}(x) < \widetilde{a_1 \wedge a_2}(x), \quad \text{all } x \in F.$$

The set $U = \{x \in K \mid \hat{g}(x) < \widetilde{a_1 \wedge a_2}(x)\}$ is open and contains F. By compactness there is an α such that $F_\alpha \subset U$, and by Hahn-Banach separation in product space there is an $a \in A(F_\alpha)$ such that

$$\hat{g}|F_\alpha < a < \widetilde{a_1 \wedge a_2}|F_\alpha.$$

Since the inequalities above are strict and the functions are semi-continuous in the appropriate sense, we may interpolate an $\varepsilon > 0$, so as to render Theorem II.6.15 appliciable. Hence there exists a $b \in A(K)$ such that

$$b|F = a|F, \qquad 0 < b < a_1, a_1.$$

Since $b|F = a|F > \hat{g}|F$, this proves the directedness, and we are through. □

It follows by Proposition II.6.21 that

$$\{F \cap \partial_e K \mid F \text{ closed split face of } K\} \tag{6.49}$$

is closed under arbitrary intersections and finite unions. In fact, the first statement is completely trivial, and the second follows from the identity

$$(F_1 \cap \partial_e K) \cup (F_2 \cap \partial_e K) = \operatorname{conv}(F_1 \cup F_2) \cap \partial_e K.$$

Hence (6.49) is the collection of closed sets for a topology, which we shall call the *facial topology* of $\partial_e K$.

Clearly the facial topology of $\partial_e K$ is weaker (coarser) than the relative topology (induced from the given topology of K).

Note that the facial topology may very well be trivial. For example, the only convex compact subsets of $\mathbb{R}^3$ with non-trivial facial topology, are the n-simplexes for $n \leq 3$, and the convex hull of a plane convex set with a point off its plane.

Proposition II.6.21. *$\partial_e K$ is compact (but possibly non-Hausdorff) in the facial topology.*

Proof. Let $\{S_\alpha\}$ be a collection of subsets of $\partial_e K$ which are closed in the facial topology, say $S_\alpha = F_\alpha \cap \partial_e K$ where each F_α is a closed split face.

If $\{S_\alpha\}$ has the finite intersection property, then $\{F_\alpha\}$ has, and it follows from the compactness of K that $\bigcap_\alpha F_\alpha \neq \emptyset$. Since $\bigcap_\alpha F_\alpha$ is a closed face, it meets $\partial_e K$, and so

$$\bigcap_\alpha S_\alpha = \left(\bigcap_\alpha F_\alpha\right) \cap \partial_e K \neq \emptyset.$$

The proof is complete. □

This may be the appropriate time to see what happens for simplexes.

Theorem II.6.22. *If K is a simplex, then every closed face F of K is split.*

Proof. Let b be a real valued function on K such that $b|F \in A(F)^+$ and $b|(K \backslash F) = 0$. Now b is u.s.c. and convex and it follows by Theorem II.3.8 that $\hat{b}$ is affine. This completes the proof by virtue of Theorem II.6.18. □

Remark. The brevity and elegance of the above proof is bought at the expence of all the hard work leading op to Theorem II.6.18, and the reader may find it instructive to give an independent proof based directly on the definition of a split face. For example, one can use the fact that $\tilde{K}$ is a (locally compact) lattice cone to obtain a decomposition:

$$x = \lambda y + (1-\lambda) z, \quad y \in F, \quad z \in F', \quad 0 \leqq \lambda \leqq 1,$$

where $\lambda y = \sup\{u \in \tilde{F} \mid u \leqq x\}$; and then show uniqueness by proving $\operatorname{lin} F \cap \operatorname{lin} F' = \{0\}$ (cf. Prop. II.6.1). This is essentially the same as the standard proof of Riesz' decomposition of a complete vector lattice into a direct (ordered) sum of an (order-)closed lattice ideal and its complementary ideal. (Closed lattice ideal = "bande" in Bourbaki's terminology.) [342], [87].

It follows from Theorem II.6.22 that the closed sets of the facial topology of the extreme boundary of a *simplex* K are *all* sets $F \cap \partial_e K$ where F is a closed face of K. In particular, the facial topology of the extreme boundary of a simplex K (consisting of more than one point) can never be trivial, since $\operatorname{conv}(x_1, \ldots, x_n)$ is a closed face for any finite set $\{x_1, \ldots, x_n\}$ of extreme points. However, the facial topology may well be non-Hausdorff, even for simplexes. In the next section we shall show that the facial topology is Hausdorff for Bauer simplexes and no other convex compact sets.

*We have seen that closed split faces are Archimedean. Their annihilators are Archimedean ideals in $A(K)$; hence they are kernels of certain order homomorphisms. We proceed to give a more explicite characterization of these order homomorphisms, and we shall see that they generalize in a natural way the linear lattice homomorphisms of vector lattices.

Recall that for any $a \in A(K)^+$ the (order-)*interval* $[0, a]$ is defined by:

$$[0, a] = \{b \in A(K) \mid 0 \leqq b \leqq a\}. \tag{6.50}$$

Clearly, if φ is an order preserving map, then for $a_1, a_2 \geqq 0$:

$$\varphi([0, a_1] \cap [0, a_2]) \subset [0, \varphi(a_1)] \cap [0, \varphi(a_2)]. \tag{6.51}$$

We shall see that the order homomorphisms we are looking for, are those for which the inclusion (6.51) is an equality "up to an ε". Specifically, we shall have the following:

Proposition II.6.23. *Let J be an Archimedean ideal in $A(K)$ and let $\varphi: A(K) \to A(K)/J$ be the canonical homomorphism. The face $F = J^\perp$ is split iff for every $\varepsilon > 0$ and any two $a_1, a_2 \in A(K)^+$:*

$$[0, \varphi(a_1)] \cap [0, \varphi(a_2)] \subset \varphi([0, a_1 + \varepsilon] \cap [0, a_2 + \varepsilon]) \tag{6.52}$$

Proof. Formula (6.52) means that for every $a_0 \in [0, \varphi(a_1)] \cap [0, \varphi(a_2)]$, i.e. every $a_0 \in A(F)^+$ such that $a_0 \leqq a_1|F, a_2|F$, there exists a $c \in A(K)^+$ such that $c|F = a_0$, $c \leqq a_1 + \varepsilon$ and $c \leqq a_2 + \varepsilon$. Using Theorem II.6.15 and Theorem II.6.18, we can prove that this statement is equivalent to F being split. □

If $A(K)$ is a lattice, i.e. if K is a Bauer simplex (Th. II.4.1), then the relation (6.52) means that φ is a lattice homomorphism, and in this case one can take $\varepsilon = 0$. In general one can not take $\varepsilon = 0$ in formula (6.52) of the above proposition. An example to this effect has been given by Stefánsson in connection with an application to C^*-algebras [373].

We shall say that an Archimedean ideal J of $A(K)$ is a *near-lattice ideal* if the corresponding homomorphism $\varphi: A(K) \to A(K)/J$ satisfies (6.52). Then we can phrase Proposition II.6.23 as follows: *The map $J \to J^\perp$ is a bijection of the set of near-lattice ideals of $A(K)$ onto the set of closed split faces of K.*

For every $x \in \partial_e K$ we denote by F_x the smallest closed split face containing x. (It is well defined by Proposition II.6.20) Also we shall say that an ideal J of $A(K)$ is *fixed* at the point $x \in \partial_e K$ if $a(x)=0$ for all $a \in J$. By the complete duality of near lattice ideals and closed split faces we have the following:

Proposition II.6.24. *For every $x \in \partial_e K$ there exists a largest near-lattice ideal J_x which is fixed at x, and J_x is given by the explicite formula:*

$$J_x = \{a \in A(K) \mid a(y)=0 \text{ for all } y \in F_x\} \tag{6.53}$$

The ideal J_x defined in (6.53), is called the *primitive ideal* at x, and $\{J_x \mid x \in \partial_e K\}$ is called the *primitive ideal space* of $A(K)$.

Note that J_x is not always a maximal near-lattice ideal; it is only maximal among those fixed at x. As an example one may consider the closed convex hull K of a regular triangle D and a point p off the plane of the triangle. If x is the center of D, then F_x is seen to be all of D, and it contains properly every $F_y = \{y\}$ with $y \in \partial_e D$. Hence J_x is a proper subset of J_y in this case.

We shall see that for a large class of compact convex sets K the facial topology of $\partial_e K$ is closely related to a "hull kernel" topology of the primitive ideal space. The relevant condition is *Størmer's axiom:*

$$\text{All } F_\alpha \text{ closed split faces} \Rightarrow \text{cl.conv}\left(\bigcup_\alpha F_\alpha\right) \text{ split face}. \tag{6.54}$$

Lemma II.6.25. *If K satisfies Størmer's axiom, then the set $\mathscr{E} = \{F_x \mid x \in \partial_e K\}$ can be topologized by the closure operation $\mathscr{S} \rightsquigarrow \bar{\mathscr{S}}$, where*

$$\bar{\mathscr{S}} = \left\{F \in \mathscr{E} \,\middle|\, F \subset \text{cl.conv}\left(\bigcup_{G \in \mathscr{S}} G\right)\right\}, \tag{6.55}$$

and the map $x \rightsquigarrow F_x$ will be continuous and open from the facial topology of $\partial_e K$ to this topology of $\mathscr{E}$. Moreover, the topology of $\mathscr{E}$ is T_0.

Proof. 1. Clearly $\bar{\emptyset} = \emptyset$, $\bar{\mathscr{E}} = \mathscr{E}$, and $\mathscr{S} \subset \bar{\mathscr{S}}$ for all $\mathscr{S} \subset \mathscr{E}$. If $H \in \bar{\mathscr{S}}$, then by definition $H \subset \text{cl.conv}\left(\bigcup_{G \in \mathscr{S}} G\right)$, and so

$$\bigcup_{H \in \mathscr{S}} H \subset \text{cl.conv}\left(\bigcup_{G \in \mathscr{S}} G\right).$$

From this it follows that

$$\text{cl.conv}\left(\bigcup_{H \in \bar{\mathscr{S}}} H\right) \subset \text{cl.conv}\left(\bigcup_{G \in \mathscr{S}} G\right),$$

and this implies $\bar{\bar{\mathscr{S}}} = \bar{\mathscr{S}}$.

It remains to prove that $\overline{\mathscr{S}\cup\mathscr{T}}=\bar{\mathscr{S}}\cup\bar{\mathscr{T}}$. The relation $\bar{\mathscr{S}}\cup\bar{\mathscr{T}}\subset\overline{\mathscr{S}\cup\mathscr{T}}$ is trivial, and to prove the opposite relation we consider an element of $\overline{\mathscr{S}\cup\mathscr{T}}$, say $F_x\in\overline{\mathscr{S}\cup\mathscr{T}}$ where $x\in\partial_e K$. Now

$$F_x\subset \mathrm{cl.conv}\Big(\bigcup_{G\in\mathscr{S}\cup\mathscr{T}} G\Big),$$

and so

$$x\in \mathrm{cl.conv}\Big(\bigcup_{G\in\mathscr{S}\cup\mathscr{T}} G\Big).$$

Since $x\in\partial_e K$, we can apply Milman's theorem to obtain

$$x\in\overline{\bigcup_{G\in\mathscr{S}\cup\mathscr{T}} G}=\Big(\overline{\bigcup_{G\in\mathscr{S}} G}\Big)\cup\Big(\overline{\bigcup_{G\in\mathscr{T}} G}\Big).$$

By symmetry we may assume

$$x\in\overline{\bigcup_{G\in\mathscr{S}} G}\subset \mathrm{cl.conv}\Big(\bigcup_{G\in\mathscr{S}} G\Big).$$

By Størmer's axiom the right-hand term of this relation is a closed split face. Hence:

$$F_x\subset \mathrm{cl.conv}\Big(\bigcup_{G\in\mathscr{S}} G\Big)=\bar{\mathscr{S}}. \tag{6.56}$$

This completes the proof of the relation $\overline{\mathscr{S}\cup\mathscr{T}}\subset\bar{\mathscr{S}}\cup\bar{\mathscr{T}}$. Thus we have proved that (6.55) defines a topological closure operation on $\mathscr{E}$.

2. Let $\mathscr{S}$ be a closed subset of $\mathscr{E}$ and consider the set $S=\{x\in\partial_e K\,|\,F_x\in\mathscr{S}\}$.

We shall have

$$F_x\in\mathscr{S} \;\Leftrightarrow\; F_x\subset \mathrm{cl.conv}\Big(\bigcup_{G\in\mathscr{S}} G\Big) \;\Leftrightarrow\; x\in \mathrm{cl.conv}\Big(\bigcup_{G\in\mathscr{S}} G\Big),$$

where the non-trivial part of the last equivalence follows by Størmer's axiom as in the proof of formula (6.56) above.

It now follows that

$$S=\partial_e K\cap \mathrm{cl.conv}\Big(\bigcup_{G\in\mathscr{S}} G\Big),$$

and the right hand term of this equation is a closed subset of $\partial_e K$ in the facial topology. This proves the continuity of the map $x \rightsquigarrow F_x$.

To prove that $x\rightsquigarrow F_x$ is an open map, we consider a subset U of $\partial_e K$ which is open in the facial topology, say $U=\partial_e K\backslash F$ where F is a closed split face, and we shall show that $\mathscr{U}=\{F_x\,|\,x\in U\}$ is an open subset of $\mathscr{E}$.

In this connection we shall first establish the formula

$$\mathrm{cl.conv}\Big(\bigcup_{H\in\mathscr{E}\backslash\mathscr{U}} H\Big)\subset F. \tag{6.57}$$

In fact if $H\in\mathscr{E}\backslash\mathscr{U}$, then $H=F_y$ for some $y\in\partial_e K$, and we must have $y\in F$ for otherwise $y\in U$ and then $H=F_y\in\mathscr{U}$ contrary to assumption. Since $y\in F$, we shall have $H=F_y\subset F$, and (6.57) follows.

Next we observe that it follows from the definition of the topology of $\mathscr{E}$ that the following formula is a necessary and sufficient condition that a subset $\mathscr{U}$ of $\mathscr{E}$ be open:

$$G\in\mathscr{U} \;\Rightarrow\; G\not\subset \mathrm{cl.conv}\Big(\bigcup_{H\in\mathscr{E}\backslash\mathscr{U}} H\Big)\,. \tag{6.58}$$

Now we shall prove that (6.58) is valid in the present situation, and we consider a $G\in\mathscr{U}$. By the definition of $\mathscr{U}$, we have $G=F_x$ for some $x\in U$. Now $x\notin F$, and it follows from (6.57) that

$$x\notin \mathrm{cl.conv}\Big(\bigcup_{H\in\mathscr{E}\backslash\mathscr{U}} H\Big)\,.$$

Hence, the implication (6.58) is valid, and we are through.

3. To prove that the topology of $\mathscr{E}$ is T_0, we consider two distinct elements F, G of $\mathscr{E}$, say $F\not\subset G$. By definition of closure $\overline{\{G\}}=\{H\in\mathscr{E}\,|\,H\subset G\}$, and it follows that $F\notin\overline{\{G\}}$. Hence $\mathscr{E}\backslash\overline{\{G\}}$ is an open neighbourhood of F which excludes G, and the proof is complete. ▯

Lemma II.6.26. *If K satisfies Størmer's axiom and $\{F_\alpha\}$ is a family of closed split faces, then*

$$\mathrm{cl.conv}\Big(\bigcup_\alpha F_\alpha\Big) = \Big(\bigcap_\alpha F_\alpha^\perp\Big)^\perp, \tag{6.59}$$

$$\Big(\mathrm{cl.conv}\Big(\bigcup_\alpha F_\alpha\Big)\Big)^\perp = \bigcap_\alpha F_\alpha^\perp. \tag{6.60}$$

Proof. 1. It is a simple verification to show that

$$\mathrm{cl.conv}\Big(\bigcup_\alpha F_\alpha\Big) \subset \Big(\bigcap_\alpha F_\alpha^\perp\Big)^\perp.$$

To prove the opposite relation, we consider an $x\in K\backslash\mathrm{cl.conv}\Big(\bigcup_\alpha F_\alpha\Big)$. By Hahn-Banach separation there is an $a\in A(K)$ such that $a(x)>0$ and $a<0$ on $\mathrm{cl.conv}\Big(\bigcup_\alpha F_\alpha\Big)$. By Theorem II.6.15 there exists a $b\in A(K)$ such that $b\geqq a$ and $b=0$ on $\mathrm{cl.conv}\Big(\bigcup_\alpha F_\alpha\Big)$. Now $b\in\bigcap_\alpha F_\alpha^\perp$ and $b(x)>0$, and so

$$x\notin\Big(\bigcap_\alpha F_\alpha^\perp\Big)^\perp.$$

2. The verification of (6.60) is straightforward. ▯

We shall denote the primitive ideal space of $A(K)$ by the symbol $\operatorname{Prim} A(K)$. For any set $\mathscr{J} \subset \operatorname{Prim} A(K)$ we define its *kernel* to be the set $k(\mathscr{J}) \subset A(K)$ defined by:

$$k(\mathscr{J}) = \bigcap_{J \in \mathscr{J}} J. \tag{6.61}$$

If K satisfies Størmer's axiom, then $k(\mathscr{J})$ is a near-lattice ideal by formula (6.60) of Lemma II.6.26.

For every near-lattice ideal I of $A(K)$ we define its *hull* to be the set $h(I) \subset \operatorname{Prim} A(K)$ defined by:

$$h(I) = \{J \in \operatorname{Prim} A(K) \mid I \subset J\}. \tag{6.62}$$

Lemma II.6.27. *If K satisfies Størmer's axiom, then $F \to F^{\perp}$ is a homeomorphism of the set $\mathscr{E} = \{F_x \mid x \in \partial_e K\}$ endowed with the topology of Lemma* II. 6.25, *onto the set* $\operatorname{Prim} A(K)$ *endowed with a topology with closure operation*:

$$\mathscr{J} \rightsquigarrow h(k(\mathscr{J})). \tag{6.63}$$

Proof. Let $\mathscr{S} \subset \mathscr{E}$ and define $\mathscr{J} = \{F^{\perp} \mid F \in \mathscr{S}\}$. By formula (6.59) we shall have:

$$\operatorname{cl.conv}\Big(\bigcup_{F \in \mathscr{S}} F\Big) = \Big(\bigcap_{F \in \mathscr{S}} F^{\perp}\Big)^{\perp} = \Big(\bigcap_{J \in \mathscr{J}} J\Big)^{\perp} = k(\mathscr{J})^{\perp}.$$

Hence we obtain the following chain of equivalences where $G \in \mathscr{E}$:

$$\begin{aligned} G \in \overline{\mathscr{S}} &\Leftrightarrow G \subset \operatorname{cl.conv}\Big(\bigcup_{F \in \mathscr{S}} F\Big) \\ &\Leftrightarrow G \subset k(\mathscr{J})^{\perp} \\ &\Leftrightarrow k(\mathscr{J}) \subset G^{\perp} \\ &\Leftrightarrow G^{\perp} \in h(k(\mathscr{J})). \end{aligned}$$

In other words, the map $F \rightsquigarrow F^{\perp}$ carries the closure operation of $\mathscr{E}$ into the operation defined by formula (6.63). This completes the proof. $\square$

The topology defined by (6.63) is called the *hull-kernel* topology (or the *Jacobson topology*) of $\operatorname{Prim} A(K)$. (Note that it is defined only if K satisfies Størmer's axiom.)

Summing up the information of Lemma II.6.25 and Lemma II.6.27, we can state:

Theorem II.6.28. *If K satisfies Størmer's axiom, then the hull-kernel topology of* $\mathrm{Prim}\,A(K)$ *is* T_0, *and the map* $x \rightsquigarrow J_x$ *is continuous and open from the facial topology of* $\partial_e K$ *to the hull-kernel topology of* $\mathrm{Prim}\,A(K)$.

A continuous affine map $\alpha: K \to K$ which is a bijection and satisfies $\alpha(F) = F$ for all $F \in \mathscr{E}$, will be called an *inner automorphism* of K. The group of inner automorphisms will be denoted by U, and we shall write $x \sim y \pmod U$ if $y = \alpha(x)$ for some $\alpha \in U$. Clearly for $x, y \in \partial_e K$:

$$x \sim y \pmod U \;\Rightarrow\; F_x = F_y . \tag{6.64}$$

We shall say that K admits *sufficiently many* inner automorphisms if for every $x \in \partial_e K$:

$$F_x = \mathrm{cl.conv}\,\{\alpha(x) \mid \alpha \in U\} . \tag{6.65}$$

Lemma II.6.29. *If K is a compact convex set satisfying Størmer's axiom and admitting sufficiently many inner automorphisms, then $\varphi: x \rightsquigarrow F_x$ is an open map from $\partial_e K$ endowed with the relative topology onto the set $\mathscr{E}$ endowed with the topology defined in Lemma* II.6.25.

Proof. We consider an arbitrary open set V in K, and we shall prove that the set $\mathscr{V} = \varphi(V \cap \partial_e K)$ is open in $\mathscr{E}$ by using the general criterion (6.58).

Thus, we assume for contradiction that there exists a $G \in \mathscr{V}$ such that

$$G \subset \mathrm{cl.conv}\left(\bigcup_{H \in \mathscr{E} \setminus \mathscr{V}} H \right) . \tag{6.66}$$

Since the members of $\mathscr{E}$ are *faces* of K, their extreme points will belong to $\partial_e K$, and so we may convert (6.66) into

$$G \subset \mathrm{cl.conv}\left(\partial_e K \cap \bigcup_{H \in \mathscr{E} \setminus \mathscr{V}} H \right) . \tag{6.67}$$

Let $G = F_x$ where $x \in \partial_e K \cap V$, and apply Milman's theorem to obtain:

$$x \in \overline{\partial_e K \cap \bigcup_{H \in \mathscr{E} \setminus \mathscr{V}} H} . \tag{6.68}$$

The set V is an open neighbourhood of x in K, and by (6.68) there exists a point x' such that

$$x' \in V \cap \partial_e K \cap \bigcup_{H \in \mathscr{E} \setminus \mathscr{V}} H .$$

Now we choose $y \in \partial_e K$ such that

$$x' \in V \cap \partial_e K \cap F_y , \tag{6.69}$$

and such that $F_y \in \mathscr{E} \backslash \mathscr{V}$, or equivalently:

$$F_y \neq F_z, \quad \text{for all } z \in V \cap \partial_e K. \tag{6.70}$$

It follows from (6.69) and the existence of sufficiently many inner automorphisms that

$$x' \in \text{cl.conv}\{\alpha(y) \mid \alpha \in U\},$$

and by Milman's theorem we obtain

$$x' \in \overline{\{\alpha(y) \mid \alpha \in U\}}.$$

Since V is an open neighbourhood of x', there exists a point z_0 such that

$$z_0 \in V \cap \{\alpha(y) \mid \alpha \in U\}.$$

Now $z \in \partial_e K$ since every $\alpha \in U$ maps $\partial_e K$ onto itself. By definition, $z_0 \in \{\alpha(y) \mid \alpha \in U\}$ means that $z_0 \sim y \pmod U$. Hence we have

$$z_0 \sim y \pmod U, \qquad z_0 \in V \cap \partial_e K. \tag{6.71}$$

This contradicts (6.70) by virtue of (6.64), and the proof is complete. □

Theorem II.6.30. *If K is a compact convex set satisfying Størmer's axiom and admitting sufficiently many inner automorphisms, then the map $x \rightsquigarrow J_x$ is continuous and open from the relative topology of $\partial_e K$ to the hull-kernel topology of* Prim $A(K)$.

Proof. Continuity follows from Theorem II.6.28, since the relative topology is stronger (finer) than the facial topology of $\partial_e K$, and openness follows by Lemma II.6.27 and Lemma II.6.29. □

Corollary II.6.31. *If K is a compact convex set satisfying Størmer's axiom and admitting sufficiently many inner automorphisms, then* Prim $A(K)$ *is a Baire space in the hull-kernel topology.*

Proof. Straightforward application of Corollary I.5.14 and Theorem II.6.30 above. □

Remark. It is beyond the scope of this book to treat the various applications of compact convex sets to analysis. Nevertheless, we feel that a few words should be said at this point about the connections with C^*-algebras, which have so strongly influenced the development of the material in the present section. If $\mathscr{A}$ is a C^*-algebra with identity

I and $\mathscr{A}_{sa}$ is the self-adjoint part of $\mathscr{A}$, then $(\mathscr{A}_{sa}, I)$ is an order-unit space with a state space S consisting of all $p \in \mathscr{A}^*$ such that $p(I)=1$ and $p(a^* a) \geqq 0$ for all $a \in \mathscr{A}$. The Archimedean (order-)ideals of $\mathscr{A}_{sa}$ are exactly the self-adjoint parts of the closed two-sided ideals of the algebra $\mathscr{A}$; and they are automatically near-lattice ideals. Dually, the Archimedean faces of S are exactly those faces F which are "invariant" in that

$$p \in F, \quad a \in \mathscr{A}, \quad p(a^* a) \neq 0 \Rightarrow p_a \in F,$$

where $p_a(x) = p(a^* a)^{-1} p(a^* x a)$; and these faces are automatically split. If π_p is the cyclic representation of $\mathscr{A}$ corresponding to a state p, then J_p is the self-adjoint part of ker π_p. Hence the primitive ideal space of convexity theory can be identified with the customary primitive ideal space of C^*-theory. The validity of Størmer's axiom is an easy consequence of the equality $J_p = (\ker \pi_p)_{sa}$, but the existence of sufficiently many inner automorphisms is more profound, resting on Kadison's transitivity theorem.—The reader may consult [13] for proofs, and also for relevant references to the literature of C^*-algebras.

Notes. The notion of a split face is quite recent. Together with the dual notion of a near-lattice ideal it was introduced by Alfsen and Bai Andersen in 1969 [13] and independently by Perdrizet at about the same time [320], [321]. (Cf. also Combes and Perdrizet [113], [114].) However, much of the material of the present section can be traced back before that time.—In 1965 Alfsen proved that every closed face of a simplex is "split" [7], but this result was already implicit in Hustad's work from 1962 on the existence of "suplementary subcones" in a lattice cone [219], and that in turn is related to Riesz' classical theorem from 1939 on the decomposition of a complete vector lattice into a direct ordered sum of complementary closed lattice ideals ("bandes" in Bourbaki's terminology) [342].—The importance of extension theorems like Theorem II.6.15 was recognized by Effros in his 1967 paper on "Structure in simplexes" [155]. He proved it with $\varepsilon=0$ for a closed face of a simplex, and used this result as one of the main tools in his subsequent investigations.—The general Theorem II.6.15 was proved in 1969 by Alfsen and Andersen [13]; and in this case the proof is no more translation of the simplex proof which is rendered inapplicable by the absence of Edwards' separation theorem (Th. II.3.10). One has to use an independent 2^{-n}-argument based on the preceding results (Prop. II.6.4 and Lem. II.6.13).—It was shown in 1969 by Stefánsson that it is impossible to take $\varepsilon=0$ in Theorem II.6.15 [373]. Stefánssons example is taken from the theory of operator algebras, and it is of interest in itself, answering a question of Dixmier about ideals of a von Neumann algebra.—The concept of a facial topology was first introduced for simplexes by Effros in 1967 [155].—A similar topology was defined for a class of (not necessarily simplicial) compact convex sets K by Størmer in 1968 [382]. Størmer's axioms are formulated as conditions on $A(K)$; which is said to be *smooth* when they are satisfied. The axioms are chosen such that a facial topology can be defined by means of *all* Archimedean faces, and such that they shall be enjoyed by C^*-algebras. The condition which we have called "Størmer's axiom" is just one of the axioms for smooth ordered spaces, but it is a very essential one, and in fact the only one which is not automatically satisfied when we restrict ourself to use split faces in the defintion of the facial topology.—The notion of a "compact convex set with sufficiently many inner automorphisms" is due to Alfsen and Andersen [13]. So are Theorems II.6.28 and II.6.30, although the method of proof is strongly influenced by Størmer's structure theory for smooth ordered spaces [382]. Theorem II.6.30 was originally proved for C^*-algebras in 1961 by Glimm [187] (see also Fell [178]). Note also that some of the background of Corollary II.6.31 has been given in the historical notes to Ch. I, § 5.

§ 7. The Concept of Center for $A(K)$

Again we shall assume that K is a fixed compact convex set, regularly embedded in $E(\cong A(K)^*)$.

We start by the following lemma of Bauer [41]:

Lemma II.7.1. *Let* $a, b: K \to \mathbb{R}$ *be two bounded u.s.c. functions, and assume that a is affine and b is concave. If* $a|\partial_e K \leqq b|\partial_e K$, *then* $a \leqq b$.

Proof. We assume for contradiction that there is a point $x \in K$ such that $b(x) < a(x)$. By Hahn-Banach separation in product space there exists a $c \in A(K)$ such that $b < c$ and $c(x) < a(x)$. (For details cf. the proof of Prop. II.1.2). The function $a - c$ is affine and u.s.c., and so it will attain its maximum at a point $y \in \partial_e K$.

Now

$$a(y) - c(y) \leqq a(y) - b(y) \leqq 0,$$

and by assumption

$$0 < a(x) - c(x) \leqq a(y) - c(y).$$

This contradiction completes the proof. □

Theorem II.7.2. *If* $f:\partial_e K \to \mathbb{R}$ *is bounded and u.s.c. in the facial topology and if* $a \in A(K)^+$, *then there exists a unique u.s.c. affine function* $c: K \to \mathbb{R}$ *such that:*

$$c(x) = f(x)a(x), \text{ for all } x \in \partial_e K. \tag{7.1}$$

Proof. 1. We assume first that $f(\partial_e K) \subset [0,1]$, and we shall construct a descending sequence $\{c_n\}$ of u.s.c. affine functions which converges to $f \cdot (a|\partial_e K)$ on $\partial_e K$.

For every natural number n we define

$$C_{n,i} = \{x \in \partial_e K \mid f(x) \geqq 2^{-n} i\}, \qquad i = 0, 1, \ldots, 2^n.$$

By hypothesis every $C_{n,i}$ is closed in the facial topology. Hence there exists a closed split face $F_{n,i}$ such that $C_{n,i} = \partial_e K \cap F_{n,i}$. We define a real valued function $b_{n,i}$ by writing $b_{n,i}(x) = 2^{-n} a(x)$ if $x \in F_{n,i}$ and $b_{n,i}(x) = 0$ if $x \in K \backslash F_{n,i}$. Then we consider the function $a_{n,i} = \hat{b}_{n,i}$. By Theorem II.6.18 $a_{n,i}$ is an affine function, and by the upper semi-continuity of $b_{n,i}$ we obtain (cf. Prop. I.4.1):

$$a_{n,i}(x) = \begin{cases} 2^{-n} a(x) & \text{for} \quad x \in C_{n,i}, \\ 0 & \text{for} \quad x \in \partial_e K \backslash C_{n,i}. \end{cases}$$

Now we define upper semi-continuous affine functions:

$$c_n = \sum_{i=0}^{2^n} a_{n,i}, \qquad n = 1, 2, \ldots. \tag{7.2}$$

It is a routine verification to prove that $\{c_n | \partial_e K\}$ is a descending sequence of step-functions which converges (even uniformly) to $f \cdot (a|\partial_e K)$. It follows by Lemma II.7.1 that $\{c_n\}$ is a descending sequence as well. Hence $c = \inf_n c_n$ is an upper semi-continuous affine function on K which satisfies (7.1). Hence we have proved the existence in case $f(\partial_e K) \subset [0,1]$.

2. Assume next that $-\lambda \leqq f \leqq \lambda$ where $\lambda \in \mathbb{R}^+$. The function

$$f' = (2\lambda)^{-1}(f + \lambda)$$

is facially continuous and takes values in $[0,1]$. By the first part of the proof there is an u.s.c. affine function c' on K such that

$$c'(x) = f'(x) \cdot a(x), \quad \text{for all } x \in \partial_e K.$$

It follows that $c = 2\lambda c' - \lambda a$ satisfies (7.1). Hence we have proved the existence in full generality.

3. It follows from Lemma II.7.1 that if two u.s.c. affine functions coincide on $\partial_e K$, then they are equal. This gives the uniqueness, and we are through. □

Corollary II.7.3. *With the assumptions and notations of Theorem* II.7.2 *we shall have*

$$c=\widehat{f\cdot(a|\partial_e K)}. \tag{7.3}$$

Proof. By Lemma II.7.1, $c\leqq g$ for every u.s.c. concave function g such that $g\geqq f\cdot(a|\partial_e K)$. From this (7.3) follows. □

Corollary II.7.4. *If* $f:\partial_e K\to\mathbb{R}$ *is continuous in the facial topology and if* $a\in A(K)$, *then there exists a unique function* $c\in A(K)$ *such that*

$$c(x)=f(x)a(x), \quad \text{for all } x\in\partial_e K. \tag{7.4}$$

Moreover:

$$\|c\|=\sup_{x\in\partial_e K}|f(x)a(x)|. \tag{7.5}$$

Proof. Let $-\lambda\leqq a\leqq\lambda$. By Theorem II.7.2 there exist two u.s.c. affine functions c_1,c_2 on K such that for every $x\in\partial_e K$:

$$c_1(x)=f(x)(a(x)+\lambda), \qquad c_2(x)=-f(x).$$

The function $c=c_1+\lambda c_2$ is u.s.c. and affine, and it satisfies:

$$c(x)=f(x)a(x) \quad \text{for all } x\in\partial_e K.$$

By the same argument as in the proof of Corollary II.7.3, we conclude that

$$c=\widehat{f\cdot(a|\partial_e K)}.$$

Hence we have proved that $\widehat{f\cdot(a|\partial_e K)}$ is affine. Reversing signs, we conclude that the function

$$d=\widetilde{f\cdot(a|\partial_e K)}$$

is affine as well.

Now $c-d$ is an u.s.c., affine and positive function. It attains its maximum on $\partial_e K$, and since $c|\partial_e K=d|\partial_e K$ we must have $c=d$. It follows that c is a continuous affine function satisfying (7.4).

The uniqueness is trivial, and so is formula (7.5). □

Taking $a=1_K$ in Corollary II.7.4 we obtain the following extension theorem for facially continuous functions:

Corollary II.7.5. *If* $f:\partial_e K\to\mathbb{R}$ *is continuous in the facial topology, then f can be (uniquely) extended to a continuous affine function on K.*

It is of some interest to characterize those functions in $A(K)$ whose restrictions to $\partial_e K$ are continuous in the facial topology. The answer to this problem invokes the notion of "center" which will be defined in the sequel. However, we shall first give some applications of Corollary II.7.5 which are not directly connected with the construction of the center of $A(K)$.

Theorem II.7.6. *The facial topology and the relative topology of $\partial_e K$ will coincide iff K is a Bauer simplex.*

Proof. 1. Let K be a Bauer simplex and consider a subset G of $\partial_e K$ which is closed in the relative topology. Since $\partial_e K$ is closed, G is a compact subset of K. Let F be the set of all points in K which are barycenters of probability measures supported by G. It follows by the (vague) compactness of $M_1^+(G)$ and the continuity of the barycenter map that F is compact. Also it is easily verified that F is a face and that $G = F \cap \partial_e K$. By Theorem II.6.22, F is a split face. Hence G is a closed set in the facial topology.

2. Assume next that the two topologies on $\partial_e K$ are identical. Then every function $f: \partial_e K \to \mathbb{R}$ which is continuous in the relative topology, can be extended to a continuous affine function on K (Cor. II.7.5). It follows from Theorem II.4.3 that K is a Bauer simplex, and the proof is complete. □

Lemma II.7.7. *Let K be a compact convex set, let $x \in \overline{\partial_e K} \backslash \partial_e K$, and let F_x be the smallest closed split face containing x. If u, v are two distinct extreme points of F_x and if U, V are two subsets of $\partial_e K$ which are open in the facial topology and for which $u \in U$, $v \in V$, then necessarily $U \cap V \neq \emptyset$.*

Proof. Assume for contradiction that $U \cap V = \emptyset$. Since U and V are open in the facial topology, there exist closed split faces G and H such that

$$\partial_e G = \partial_e K \backslash U, \qquad \partial_e H = \partial_e K \backslash V.$$

Now $\partial_e G \cup \partial_e H = \partial_e K$, and so

$$x \in \overline{\partial_e K} = \overline{\partial_e G} \cup \overline{\partial_e H} \subset G \cup H.$$

We may as well assume $x \in G$. It follows that $F_x \subset G$, and so $\partial_e F_x \subset \partial_e G$. This gives the contradiction

$$u \in \partial_e F_x \subset \partial_e G = \partial_e K \backslash U,$$

and the proof is complete. □

Theorem II.7.8. *The facial topology of $\partial_e K$ is Hausdorff iff K is a Bauer simplex.*

Proof. 1. If K is a Bauer simplex, then the facial topology will be equal to the relative topology (Th. II.7.6). Hence it is Hausdorff.

2. Assume next that the facial topology is Hausdorff. We claim that in this case $\partial_e K$ is a closed subset of K. To prove this claim we assume the contrary, i.e. we assume that there exists a point $x \in \overline{\partial_e K}$ which is not an extreme point. Then $F_x \neq \{x\}$, and it follows by Krein-Milman that F_x must contain two distinct extreme points u, v. By Lemma II.7.7 the points u, v can not be Hausdorff-separated in the facial topology, and we have a contradiction.

Now the facial topology is a Hausdorff topology weaker than the *compact* relative topology of $\partial_e K$, and so the two must coincide. By Theorem II.7.6 K is a Bauer simplex, and the proof is complete. □

We shall denote by $Z(A(K))$, or briefly by Z, the set of all elements $b \in A(K)$ such that for every $a \in A(K)$ there exists a $c \in A(K)$ satisfying

$$c(x) = b(x)a(x), \quad \text{for all } x \in \partial_e K. \tag{7.6}$$

The set Z will be termed the *center* of $A(K)$.

The element c of (7.6) is the (unique) extension of $b \cdot a | \partial_e K$ to a function in $A(K)$. and we may phrase the definition of Z somewhat sketchily by saying that the central elements are those which define "multiplication operators" on $A(K)$. Note, however, that the multiplication takes place on extreme points only. Formula (7.6) is inexact for $x \in K \backslash \partial_e K$.

Proposition II.7.9. *The center Z of $A(K)$ is a commutative Banach algebra and a Kakutani M-space in the ordering and norm induced from $A(K)$, and the multiplication and lattice operations are pointwise on $\partial_e K$.*

Proof. 1. Clearly Z is a linear subspace of $A(K)$. To show that Z is closed, we consider a sequence $\{b_n\}$ from Z which converges in norm to an element $b \in A(K)$. For every $a \in A(K)$ and $n = 1, 2, \ldots$ there exists a $c_n \in A(K)$ such that $c_n(x) = b_n(x)a(x)$ for all $x \in \partial_e K$. Since functions in $A(K)$ attain their maximum on $\partial_e K$, we shall have

$$\|c_n - c_m\| = \|c_n - c_m\|_{\partial_e K} \leqq \|b_n - b_m\| \cdot \|a_n\|,$$

and the right hand term tends to zero as $m, n \to \infty$. Thus $\{c_n\}$ converges to an element c of $A(K)$, and $c(x) = b(x)a(x)$ for all $x \in \partial_e K$. Hence we have proved that $b \in Z$.

2. We shall prove the following fact: For any two elements $b, b' \in Z$ there exists a (unique) element $c \in Z$ such that

$$c(x) = b(x)\,b'(x), \quad \text{for all } x \in \partial_e K. \tag{7.7}$$

By the definition of Z there exists an element c of $A(K)$ which satisfies (7.7). To show $c \in Z$, we consider an arbitrary $a \in A(K)$. Now there is an $a' \in A(K)$ such that

$$a'(x) = b'(x)\,a(x), \quad \text{for all } x \in \partial_e K.$$

Also there exists a $d \in A(K)$ such that for all $x \in \partial_e K$:

$$d(x) = b(x)\,a'(x) = b(x)\,b'(x)\,a(x) = c(x)\,a(x).$$

Hence $c \in Z$ as claimed.

Now it follows that Z is a Banach algebra, and that the multiplication is pointwise on $\partial_e K$.

3. Next we shall prove that for every $b \in Z$ there exists a $c \in Z$ such that

$$c(x) = |b(x)|, \quad \text{for all } x \in \partial_e K. \tag{7.8}$$

To this end we define $B = \{b(x)^2 \mid x \in \partial_e K\} \subset \mathbb{R}^+$, and we consider a sequence of polynomials P_n such that $P_n(\xi) \to \sqrt{\xi}$ uniformly on B. By the preceding part of the proof there exists for every $n = 1, 2, \ldots$ an element $c_n \in Z$ such that

$$c_n(x) = P_n(b(x)^2), \quad \text{for all } x \in \partial_e K.$$

Now $\{c_n\}$ is seen to be a Cauchy sequence converging to an element $c \in Z$ satisfying

$$c(x) = \sqrt{b(x)^2} = |b(x)|, \quad \text{for all } x \in \partial_e K.$$

The function $c \in Z$ satisfies (7.8), and the claim is proved.

It is easily verified that $c = \widehat{b \vee -b}$ (cf. the proof of Cor. II.7.3), and it follows that c is the least upper bound of b and $-b$ in Z (as well as in $A(K)$ and even in $-Q(K)$).

A partially ordered linear space where every pair $(b, -b)$ has a least upper bound is known to be a vector lattive, and the lattice operations are given by the familiar formulas expressing "max" and "min" by "absolute value" (cf. e.g. [61], [362]. Hence we have proved that Z is a vector lattice, and that the lattice operations are pointwise on $\partial_e K$. That Z is a Kakutani M-space, follows since the norm of Z is equal to the sup-norm on $\partial_e K$. This proof is complete. $\square$

We shall denote by $L_{ob}(A(K))$, or simply by L_{ob}, the set of all linear operators $T: A(K) \to A(K)$ which are *order bounded*, in that there exists a real number $\lambda \geqq 0$ (depending on T) such that

$$-\lambda a \leqq Ta \leqq \lambda a, \quad \text{for all } a \in A(K)^+ . \tag{7.9}$$

Clearly this is equivalent to

$$-\lambda I \leqq T \leqq \lambda I, \tag{7.10}$$

where I is the identity operator and the ordering is the natural ordering of operators on $A(K)$.

We notice that (L_{ob}, I) is an order-unit space in which the order-unit norm $\|T\|_{or}$ is the smallest positive constant λ for which (7.10) holds.

For convenience we shall also introduce the symbol $C_f(\partial_e K)$, or simply C_f, to denote the set of real valued functions on $\partial_e K$ which are continuous in the facial topology.

Our next theorem is the main result of this section. It shows that Z, C_f, and L_{ob} are essentially the same. Before giving the precise statement, we shall define the maps which will be the relevant isomorphisms:

$$\rho: Z \to C(\partial_e K), \qquad \text{defined by } b \rightsquigarrow b|\partial_e K . \tag{7.11}$$

$$\varphi: C_f \to \operatorname{Lin}(A(K), A(K)), \qquad \text{defined by } f \rightsquigarrow T_f \text{ where}$$
$$[T_f a](x) = f(x) a(x) \qquad \text{for all } a \in A(K),\ x \in \partial_e K. \tag{7.12}$$

$$\psi: L_{ob} \to A(K), \qquad \text{defined by } T \rightsquigarrow T 1_K . \tag{7.13}$$

Theorem II.7.10. *If the maps ρ, φ, ψ are defined as above, then we have a commutative diagram*

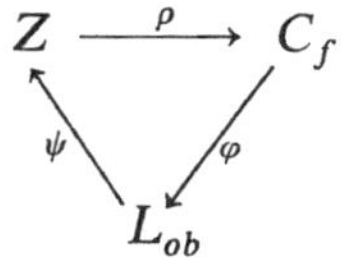

in which each map is a bijection, preserving linear structure and order structure. If Z and C_f are endowed with sup-norm, and L_{ob} is endowed with either the operator- or the order-unit-norm, then the maps are norm preserving. In particular, the two norms on L_{ob} will coincide.

Proof. 1. To prove that the restriction map ρ is a linear order-preserving map of Z into C_f, we only have to show that $b|\partial_e K$ is facially continuous for every $b \in Z$. Since Z is a linear space it suffices to prove upper semi-continuity. Hence we shall prove that for every $\alpha \in \mathbb{R}$ the set

$$C_\alpha = \{x \in \partial_e K \mid b(x) \geqq \alpha\}$$

is of the form $C_\alpha = \partial_e K \cap F$ where F is a closed split face of K. Without lack of generality we assume $0 \leqq b \leqq 1$ and $0 \leqq \alpha \leqq 1$.

Working in the multiplicative and lattice structure of Z (cf. Prop. II.7.9), we define

$$b_n = [(b + (1-\alpha)1_K) \wedge 1_K]^n \in Z, \qquad n = 1, 2, \ldots .$$

The operations in Z are pointwise on $\partial_e K$; hence

$$\lim_n b_n(x) = \begin{cases} 1 & \text{if } x \in C_\alpha, \\ 0 & \text{if } x \in \partial_e K \backslash C_\alpha. \end{cases} \tag{7.14}$$

Clearly also $\{b_n\}$ is a descending sequence on $\partial_e K$, hence on all of K.

Define a function c on K by:

$$c(x) = \inf_n b_n(x), \quad \text{for all } x \in \partial_e K .$$

Clearly c is u.s.c. and affine with values in $[0,1]$.
Hence the sets

$$F = \{x \in K \mid c(x) = 1\},$$
$$G = \{x \in K \mid c(x) = 0\},$$

are *faces*; the former closed, the latter G_δ.

It follows from (7.14) that $C_\alpha = F \cap \partial_e K$. Hence it only remains to prove that F is split.

To this end we first observe that c is an u.s.c. affine function which is equal to χ_F on $\partial_e K$ (cf. (7.14)). It follows that $c = \hat{\chi}_F$ (cf. the proof of Cor. II.7.3). By Proposition II.6.5, $F' = \hat{\chi}_F^{-1}(0) = G$ and $K = \operatorname{conv}(F \cup F')$.

To prove that K is direct convex sum of F and F' we consider an equality:

$$\lambda_1 y_1 + (1-\lambda_1) z_1 = \lambda_2 y_2 + (1-\lambda_2) z_2, \tag{7.15}$$

where $0 < \lambda_i < 1$, $y_i \in F$, $z_i \in F'$ for $i = 1, 2$.

Applying the affine function c to both sides, we obtain $\lambda_1 = \lambda_2$. Henceforth we shall denote this common constant by λ.

We assume for contradiction that $y_1 \neq y_2$. Then there is an $a \in A(K)^+$ such that $a(y_1) \neq a(y_2)$. Since $b_n \in Z$ there exist functions $d_n \in A(K)^+$ such that

$$d_n(x) = b_n(x) a(x), \quad \text{for all } x \in \partial_e K .$$

Again $\{d_n\}$ is decreasing on $\partial_e K$ and hence on all of K. We define $d = \inf_n d_n$, and conclude by the same argument as above that

$$d = \widehat{(a \cdot \chi_F) | \partial_e K}.$$

From this it follows that $d=a$ on F and $d=0$ on F' (cf. Prop. II.6.4).

Applying the affine function d to the equation (7.15), we obtain

$$\lambda a(y_1)=\lambda a(y_2).$$

Hence $a(y_1)=a(y_2)$ contrary to assumption.

2. If $f\in C_f$, then by Corollary II.7.4 there exists for every $a\in A(K)$ a function c in $A(K)$ such that

$$c(x)=f(x)a(x), \quad \text{for all } x\in\partial_e K.$$

Hence the linear operator T_f of (7.12) is well defined.

To show that T_f is order-bounded, we note that for every $a\in A(K)^+$:

$$-|f(x)|a(x)\leqq[T_f a](x)\leqq|f(x)|a(x), \quad \text{for all } x\in\partial_e K.$$

Hence

$$-\|f\|a\leqq T_f a\leqq\|f\|a, \quad \text{for all } a\in A(K)^+. \tag{7.16}$$

This shows that $T_f\in L_{ob}$ and we also note for later applications that

$$\|T_f\|_{or}\leqq\|f\|. \tag{7.17}$$

Hence $\varphi: f \rightsquigarrow T_f$ is a well defined linear map of C_f into L_{ob}. It is order preserving since

$$f\in G_f^+,\ a\in A(K)^+ \Rightarrow (T_f a)|\partial_e K\geqq 0 \Rightarrow T_f a\geqq 0.$$

3. To show that ψ is a linear and order preserving map of L_{ob} into Z, we only have to verify that $T1_K\in Z$ for every $T\in L_{ob}$. In fact we shall prove that for every $a\in A(K)$ the particular function Ta satisfies the crucial requirement:

$$[Ta](x)=[T1_K](x)a(x), \quad \text{for all } x\in\partial_e K. \tag{7.18}$$

It suffices to prove (7.18) for $0\leqq T\leqq I$. Let $x\in\partial_e K$, and assume first that

$$0<[T1_K](x)<1. \tag{7.19}$$

Now the functionals p, q on $A(K)$ defined by

$$p(a)=\frac{[Ta](x)}{[T1_K](x)},$$

$$q(a)=\frac{[(I-T)a](x)}{[(I-T)1_K](x)},$$

are seen to be *states*. Let them be fixed at points $y, z \in K$, i.e.:

$$p(a) = a(y), \quad q(a) = a(z), \quad \text{for all } a \in A(K).$$

Writing $\lambda = [T 1_K](x) \in]0,1[$, we obtain

$$a(x) = \lambda a(y) + (1-\lambda) a(z), \quad \text{for all } a \in A(K).$$

It follows that $x = \lambda y + (1-\lambda) z$, and since x is an extreme point of K and $0 < \lambda < 1$, this entails $x = y = z$.

It follows that

$$a(x) = a(y) = \frac{[T a](x)}{[T 1_K](x)}, \quad \text{for all } a \in A(K).$$

Multiplying through with $[T 1_K](x)$ we obtain the desired formula (7.18).

Next we assume that $[T 1_K](x) = 0$. Since $T \geqq 0$ it follows from the general relation

$$-\|a\| 1_K \leqq a \leqq \|a\| 1_K$$

that

$$-\|a\| T 1_K \leqq T a \leqq \|a\| T 1_K.$$

Evaluating at x, we get $[T a](x) = 0$. Hence (7.18) is proved also in this case.

Finally if $[T 1_K](x) = 1$, then we may apply a similar argument with $(I - T)$ in the place of T. (Recall that $0 \leqq T \leqq I$.)

4. By the definition of ρ, φ, ψ, and by the identity (7.18), it follows that the diagram is commutative. Consequently, all three maps are bijections.

5. It remains to verify the norm preserving nature of the maps ρ, φ, ψ. Clearly

$$\|\rho f\| = \|f\|_{\partial_e K} = \|f\|, \quad \text{for all } f \in Z. \tag{7.20}$$

By (7.17) we get

$$\|\varphi(f)\|_{or} = \|T_f\|_{or} \leqq \|f\|, \quad \text{for all } f \in C_f. \tag{7.21}$$

By the definition of order-unit norm, we shall have for every $T \in L_{ob}$: $-\|T\|_{or} I \leqq T \leqq \|T\|_{or} I$.

Hence

$$-\|T\|_{or} 1_K \leqq T 1_K \leqq \|T\|_{or} 1_K.$$

Evaluating at a point $x \in \partial_e K$ we get:

$$|[T 1_K](x)| \leqq \|T\|_{or}.$$

By (7.18) this implies for every $a \in A(K)$ and every $x \in \partial_e K$:

$$|[T a](x)| = |[T 1_K](x)| \cdot |a(x)| \leqq \|T\|_{or} \cdot \|a\|,$$

and so

$$\|Ta\| \leqq \|T\|_{or} \cdot \|a\|.$$

Hence the operator norm $\|T\|_{op}$ is related to the order-unit norm $\|T\|_{or}$ by the formula

$$\|T\|_{op} \leqq \|T\|_{or}, \quad \text{for all } T \in L_{ob}. \tag{7.22}$$

Furthermore

$$\|\psi(T)\| = \|T 1_K\| \leqq \|T\|_{op} \cdot \|1_K\| = \|T\|_{op}. \tag{7.23}$$

Combining (7.20), (7.21), (7.22), (7.23), and using the commutativity of the diagram we get a closed cycle of inequalities. This completes the proof. □

From Proposition II.7.9 and Theorem II.7.10 and by the definition of the maps ρ, φ, ψ, we immediately obtain the following:

Corollary II.7.11. *Each of the spaces Z, C_f, L_{ob} is a commutative Banach algebra and a Kakutani M-space (under the obvious multiplication, norm and ordering), and each of the maps ρ, φ, ψ is an isometry, an algebra isomorphism, and a lattice isomorphism.*

Corollary II.7.12. *The restriction to $\partial_e K$ of a function $a \in A(K)$ is continuous in the facial topology iff a is in the center of $A(K)$.*

Clearly the constant functions are in the center of $A(K)$ for every compact convex set K, and we shall say that the center of $A(K)$ is *trivial* if it consists of the constants only.

For later references we state the following:

Proposition II.7.13. *If the center of $A(K)$ is non-trivial and the set $Z_1^+ = \{a \in Z \mid 0 \leqq a \leqq 1_K\}$ is compact in some locally convex Hausdorff topology on $A(K)$, then there exists a $T \in L_{ob}$ such that*

$$T^2 = T, \quad 0 \leqq T \leqq I, \quad T \neq 0, I. \tag{7.24}$$

Proof. By the Krein-Milman theorem there exists an extreme point b of Z_1^+ such that $b \neq 0, 1_K$. We keep the notations introduced in (7.11)–(7.13) and decompose

$$\rho b = \tfrac{1}{2}(\rho b)^2 + \tfrac{1}{2}(2\rho b - (\rho b)^2) \tag{7.25}$$

It follows from the isomorphism properties of ρ (Cor. II.7.11) that ρb is an extreme point of $(C_f)_1^+ = \{g \in C_f \mid 0 \leqq g \leqq 1_K\}$. Hence the convex combination (7.25) must be trivial, and so $\rho b = (\rho b)^2$.

Now it follows from the isomorphism properties of the operator φ, that the order bounded linear operator $T=\varphi(\rho b)$ will have the desired properties (7.24). □

It is of interest to note that the center of $A(K)$ can be trivial even if K is a simplex. In this connection we shall study a special class of simplexes which are in a sense "complementary" to the Bauer simplexes.

We shall say that a simplex K is *prime* if for any two closed faces F_1, F_2 such that $K=\mathrm{conv}(F_1 \cup F_2)$, necessarily $F_1=K$ or $F_2=K$. (This definition is due to Effros and Kazdan [160].)

Observation II.7.14. *A simplex K is prime iff $U_1 \cap U_2 \neq \emptyset$ for any two non-empty sets $U_1, U_2 \subset \partial_e K$ which are open in the facial topology of $\partial_e K$.*

For the *proof* one should use the definitions of facial topology and prime simplex, and write $U_i=\partial_e K \backslash F_i$ for $i=1,2$. □

The above observation already gives some justification for the statement that the prime simplexes are "complementary" to Bauer simplexes (cf. Th. II.7.8). Another justification is obtained by passing to $A(K)$. Recall that a partially ordered vector space is said to be an *anti-lattice* if the only pairs $a_1, a_2 \in A(K)$ admitting a g.l.b., are those where either $a_1 \leqq a_2$ or $a_2 \leqq a_1$.

Theorem II.7.15. *A simplex K is prime iff $A(K)$ is an anti-lattice.*

Proof. 1. Assume first that K is prime and consider two functions $a_1, a_2 \in A(K)$ admitting a g.l.b. in $A(K)$. We denote this g.l.b. by the letter b, and we observe that it follows from the definition of envelopes (Ch. I. § 1) that

$$b=\widecheck{a_1 \wedge a_2}\,.$$

For every $x \in \partial_e K$ we shall have (by Prop. I.4.1):

$$b(x)=a_1(x) \quad \text{or} \quad b(x)=a_2(x)\,. \tag{7.26}$$

Let $c_1=a_1-b$, $c_2=a_2-b$, and define

$$F_i=\{x \in K \mid c_i(x)=0\}, \qquad i=1,2\,.$$

Clearly F_1, F_2 are closed faces, and it follows from (7.26) that

$$\partial_e K \subset F_1 \cup F_2\,.$$

It follows (by Krein-Milman) that $K \subset \operatorname{conv}(F_1 \cup F_2)$; and since K is prime, we shall have either $K = F_1$ or $K = F_2$. In the former case $a_1 = b \leqq a_2$, and in the latter case $a_2 = b \leqq a_1$. Hence we have proved that $A(K)$ is an anti-lattice.

2. Assume next that K is not prime. Then there exist two closed faces F_1, F_2 together with two points x_1, x_2 such that $K = \operatorname{conv}(F_1 \cup F_2)$ and:

$$x_1 \in \partial_e F_1 \backslash F_2, \qquad x_2 \in \partial_e F_2 \backslash F_1 .$$

It follows by application of Edwards' theorem (Th. II.3.10) with the obvious bounds, that there exist two functions $a_1, a_2 \in A(K)^+$ such that $a_1(x_1) = 1$, $a_1 | F_2 = 0$ and $a_2(x_2) = 1$, $a_2 | F_1 = 0$. Since $K = \operatorname{conv}(F_1 \cup F_2)$ the two functions a_1, a_2 will admit a g.l.b. in $A(K)$, namely the zero function, which is different from each of the two given functions. Hence $A(K)$ is not an anti-lattice. □

Observation II.7.16. *If K is a prime simplex, then the center of $A(K)$ is trivial.*

The *proof* is a direct application of Theorem II.7.10 since every function in C_f must be constant (cf. Observation II.7.14). □

Final Remark. Our definition of *center* pertains to $A(K)$-spaces. However, since every norm-complete order unit space can be (canonically) represented as an $A(K)$-space (Th. II.1.8), we can transfer the definition to norm-complete order-unit spaces in general.

*We are going to prove that there really exists a prime simplex K, and *a fortiori* a simplex K for which $A(K)$ has a trivial center.

Proposition II.7.17. *There exists a simplex K such that $\partial_e K$ with facial topology is homeomorphic to the set N of natural numbers with the weakest (coarsest) T_1-topology (whose closed sets are the finite subsets of N and N itself).*

Proof. Let $\Delta_\sigma = M_1^+(\bar{N})$ where $\bar{N}$ is the one-point compactification of the set N with discrete topology. It follows from the results of Ch. II, § 4 that Δ_σ is a Bauer simplex and that $\partial_e \Delta_\sigma \cong \bar{N}$. ($\Delta_\sigma$ is probably the simplest infinite dimensional simplex to visualize. It has countably many extreme points ε_n, $n = 1, 2, \ldots$, clustering at the extreme point ε_∞, and it is the σ-convex hull of these extreme points. Cf. the starred section at the end of Ch. I, § 4.)

Consider the (signed) measure

$$\mu = \sum_{n=1}^{\infty} 2^{-n}\varepsilon_n - \varepsilon_\infty,$$

and let $K=\varphi(\Delta_\sigma)$ where φ is the canonical mapping of $M(\bar{N})$ onto $M(\bar{N})/\mathrm{lin}(\mu)$.

We claim that φ is a bijection of $\{\varepsilon_n \mid n\in N\}$ onto $\partial_e K$.

To prove this claim, we first note the elementary fact that a subset F of K is a face iff $\varphi^{-1}(F)\cap\Delta_\sigma$ is a face of Δ_σ. From this it follows by Krein-Milman that a point x of K is an extreme point iff $\varphi^{-1}(x)\cap\Delta_\sigma$ is a singleton located on $\partial_e K$. In other words

$$\partial_e K = \{\varphi(\varepsilon_n) \mid n\in\bar{N},\ \varphi^{-1}(\varphi(\varepsilon_n))\cap\Delta_\sigma = \{\varepsilon_n\}\}. \tag{7.27}$$

Let $n\in N$ be arbitrary and consider a measure $\nu\in\varphi^{-1}(\varphi(\varepsilon_n))\cap\Delta_\sigma$. Now $\varphi(\nu)=\varphi(\varepsilon_n)$, and hence there exists a $\lambda\in\mathbb{R}$ such that

$$\nu = \varepsilon_n + \lambda\mu.$$

Evaluating the measure above on the sets $\{m\}$ with $m\in\bar{N}$, we obtain

$$\nu(\{m\}) = \begin{cases} 2^{-m}\lambda, & \text{for } m\in N, \quad m\neq n, \\ 1+2^{-n}\lambda, & \text{for } m=n, \\ -\lambda, & \text{for } m=\infty. \end{cases}$$

This is compatible with $\nu(\bar{N})=1$, $\nu\geqq 0$, only if $\lambda=0$ and $\nu=\varepsilon_n$. Hence we have shown:

$$\varphi^{-1}(\varphi(\varepsilon_n))\cap\Delta_\sigma = \{\varepsilon_n\}, \quad \text{all } n\in N. \tag{7.28}$$

Clearly $\varphi(\varepsilon_\infty)=\varphi\left(\sum_{n=1}^{\infty} 2^{-n}\varepsilon_n\right)$, and so

$$\varphi^{-1}(\varphi(\varepsilon_\infty))\cap\Delta_\sigma \neq \{\varepsilon_\infty\}. \tag{7.29}$$

Combining (7.27), (7.28), and (7.29), we conclude that φ is a bijection of N onto $\partial_e K$, as claimed.

To prove that K is a simplex, we consider a boundary affine dependence π on K, and we shall show that $\pi=0$.

The measure π must be purely atomic with point masses π_n at the extreme points $\varphi(\varepsilon_n)$ where $n\in N$. By standard duality arguments, $A(K)$ consists of all functions $a_f:\varphi(\nu)\rightsquigarrow\nu(f)$, where $\nu\in\Delta_\sigma$, $f\in C(\bar{N})$ and $\mu(f)=0$. (Cf. the proofs of Prop. I.4.15 and Prop. II.3.14 for similar arguments.) It follows that

$$0=\pi(a_f)=\sum_{n=1}^{\infty} f(n)\pi_n$$

for all $f \in C(\bar{N})$ such that $\mu(f)=0$. This implies $\sum_{n=1}^{\infty} \pi_n \varepsilon_n = \rho\mu$ for some $\rho \in \mathbb{R}$, and in fact $\rho=0$ since $\mu(\{\infty\}) \neq 0$. Hence $\pi=0$, and we have proved that K is a simplex.

It remains to prove the statement about the facial topology of $\partial_e K$, and we shall be through if we can prove that the proper closed faces of K are the sets of the form

$$F = \operatorname{conv}\{\varphi(\varepsilon_{n_k}) \mid k=1, \ldots, m\}. \tag{7.30}$$

We shall first show that a set F given by formula (7.30) is a face of K, or equivalently that $\varphi^{-1}(F) \cap \Delta_\sigma$ is a face of Δ_σ. In fact it suffices to prove

$$\varphi^{-1}(F) \cap \Delta_\sigma = \operatorname{conv}\{\varepsilon_{n_k} \mid k=1, \ldots, m\}, \tag{7.31}$$

since the set at the right hand side of (7.31) is seen to be a face of Δ_σ.

Let $v \in \varphi^{-1}(F) \cap \Delta_\sigma$, say that $\varphi(v)$ is given by a convex combination

$$\varphi(v) = \sum_{k=1}^{m} \lambda_k \varphi(\varepsilon_{n_k}).$$

It follows that there exists a $\sigma \in \mathbb{R}$ such that

$$v = \sum_{k=1}^{m} \lambda_k \varepsilon_{n_k} + \sigma\mu.$$

Now we shall have:

$$v(N \backslash \{n_1, \ldots, n_m\}) = \sigma \sum_{k=1}^{m} 2^{-k}; \qquad v(\{\infty\}) = -\sigma.$$

The first equation implies $\sigma \geqq 0$, the second implies $\sigma \leqq 0$. Hence $\sigma=0$, and so

$$v = \sum_{k=1}^{m} \lambda_k \varepsilon_{n_k},$$

as desired.

Finally we shall show that a closed face G of K which is not of the form (7.30), must be all of K. Necessarily G must have a sequence of extreme points $\varphi(\varepsilon_{n_k})$, $k=1,2,\ldots$. Now the sequence $\{\varepsilon_{n_k}\} \subset \varphi^{-1}(G) \cap \Delta_\sigma$ will converge to ε_∞. Hence $\varepsilon_\infty \in \varphi^{-1}(G) \cap \Delta_\sigma$, and so $\varphi(\varepsilon_\infty) \in G$. It follows from the definition of φ that

$$\varphi(\varepsilon_\infty) = \sum_{k=1}^{\infty} 2^{-k} \varphi(\varepsilon_k).$$

Hence for every $n \in N$ we obtain a proper convex combination

$$\varphi(\varepsilon_\infty) = 2^{-n} \varphi(\varepsilon_n) + (1-2^{-n}) \sum_{k \neq n} \frac{2^{-k}}{1-2^{-n}} \varphi(\varepsilon_k).$$

Since G is a face of K we shall have $\varphi(\varepsilon_n)\in G$ for every $n\in N$. Thus, G contains all extreme points of K, and we are through. □

It is not hard to show that the facial structure of the simplex K of Proposition II.7.17 is pathological in many ways. For example, there exist faces of K whose closures are no longer faces; and K does not enjoy Størmer's axiom. (One may consider faces defined as convex hulls of suitable sets of extreme points.) It is fairly easy to show these pathologies are impossible for Bauer simplexes, but it is less obvious that they really *do* occur for *all other* simplexes.

We shall prove this fact, and we start by a simple:

Lemma II.7.18. *The closed faces of a Bauer simplex K are exactly the sets which are representable in the form:*

$$F=\text{cl.conv.}B, \qquad B\subset\partial_e K\,. \tag{7.32}$$

Proof. 1. Let F be a closed face of K. Writing $B=F\cap\partial_e K=\partial_e F$ and using the Krein-Milman theorem we obtain (7.32). (This is of course independent of K being a Bauer simplex.)

2. Let F be defined by (7.32) and consider a convex combination

$$x=\lambda y+(1-\lambda)z\,,$$

where $y,z\in K$, $0<\lambda\leqq 1$, and $x\in F$.

Let ν and ρ be two measures in $M_1^+(K)$ which are supported by the closed set $\partial_e K$ and have barycenters y and z, respectively. Define

$$\pi=\lambda\nu+(1-\lambda)\rho\,.$$

Clearly $\pi\in M_1^+(K)$, $\text{Supp}(\mu)\subset\partial_e K$, and π has barycenter x.

Let μ be a probability measure with barycenter x which is supported by $\partial_e F$. It follows by Milman's theorem that $\partial_e F\subset\bar{B}$. Hence $\text{Supp}(\mu)\subset\bar{B}\subset\partial_e K$, and by the uniqueness property of Bauer simplexes (Th. II.4.1), $\mu=\pi$.

Since $\text{Supp}(\pi)\subset\bar{B}$ and since $\lambda>0$, we shall have $\text{Supp}(\nu)\subset\bar{B}$. Then the barycenter of ν must be in $\text{cl.conv}\,\bar{B}$, and so $y\in F$. This proves that F is a face, and we are through. □

Theorem II.7.19. *(Størmer) If K is a simplex, then the following statements are equivalent:*

(i) *K is a Bauer simplex.*

(ii) *If F is a face of K, then $\bar{F}$ is a face of K.*

(iii) *K satisfies Størmer's axiom.*

Proof. (i) $\Rightarrow$ (ii). Let $\mathscr{F}$ be the set of all $\mu \in M_1^+(K)$ with barycenter in a given face F and support contained in the closed set $\partial_e K$. Define

$$B_0 = \bigcup_{\mu \in \mathscr{F}} \operatorname{Supp}(\mu),$$

and let $B = \overline{B}_0$. Clearly $B \subset \partial_e K$, and by Lemma II.7.18 it suffices to prove that $\overline{F} = \text{cl.conv.} B$.

For every $x \in F$ there exists a $\mu \in M_1^+(K)$ with barycenter x and $\operatorname{Supp}(\mu) \subset \partial_e K$. By definition $\mu \in \mathscr{F}$, and hence $\operatorname{Supp}(\mu) \subset B$. This implies $x \in \text{cl.conv.} B$, and hence $\overline{F} \subset \text{cl.conv.} B$.

(ii) $\Rightarrow$ (iii) We consider a family $\{F_\alpha\}$ of closed faces of K (they are automatically split by Th. II.6.22), and we shall show that the set

$$F = \operatorname{conv}\left(\bigcup_\alpha F_\alpha\right)$$

is a face.

Every $x \in F$ can be written in the form $x = \sum_{i=1}^n \lambda_i x_i$ where $\sum_{i=1}^n \lambda_i = 1$, $\lambda_i > 0$, $x_i \in F_{\alpha_i}$, for $i = 1, \ldots, n$. Writing $G = \operatorname{conv}(F_{\alpha_1} \cup \cdots \cup F_{\alpha_n})$ we shall have $x \in G$.

By Corollary II.6.8, G is a (split) face of K. Hence if

$$x = \lambda y + (1-\lambda) z,$$

where $y, z \in K$, $0 < \lambda < 1$, then $y, z \in G \subset F$. Hence we have proved that F is a face of K.

By the assumption (ii) it follows that $\overline{F}$ is a face. It is split by Theorem II.6.22, and so we have proved that K satisfies Størmer's axiom.

(iii) $\Rightarrow$ (i) It follows from the validity of Størmer's axiom that the hull-kernel topology of $\operatorname{Prim} A(K)$ is well defined, and that the map $\varphi: \partial_e K \to \operatorname{Prim} A(K)$ defined by $\varphi(x) = J_x$ is continuous and open from the facial topology of $\partial_e K$ to the hull-kernel topology of $\operatorname{Prim} A(K)$ (Th. II.6.28).

Since K is a simplex, we shall have $F_x = \{x\}$ for every $x \in \partial_e K$ (Th. II.6.22). Hence φ is a *homeomorphism* of $\partial_e K$ onto $\operatorname{Prim} A(K)$.

Now we consider the following commutative diagram where $\iota: \partial_e K \to \partial_e K$ is the identity map:

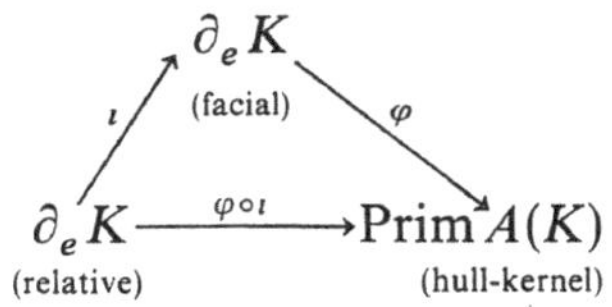

Clearly K admits sufficiently many automorphisms since every F_x is a singleton. It follows from Theorem II.6.30 that $\varphi \circ \iota$ is an open map, and hence a homeomorphism. Hence the two topologies on $\partial_e K$ must coincide, and it follows from Theorem II.7.6 that K is a Bauer simplex. The proof is complete. □

Remark. We have phrased the argument of the above proof in terms of primitive ideals in order to make easy reference to the general theorems of Ch. II. § 6. Conceptually the passage from the faces F_x to the ideals J_x is a mere complication, and the core of the argument is Lemma II.6.29. Note also that the initial requirement to simpliciality can be omitted if $\bar{F}$ is required to be a *split* face in (ii) and if the facial topology is required to be T_1 in (iii). This has recently been proved by Å. Lima (unpublished).

We close this section with a general version of the Dauns Hofmann Theorem, which is now an easy corollary of the preceding results.

Theorem II.7.20. *If K is a compact convex set satisfying Størmer's axiom, then every continuous function f on* Prim $A(K)$ *is induced by a central element of $A(K)$. Specifically, there is an isometric isomorphism $f \rightsquigarrow a_f$ of $C(\mathrm{Prim}\,A(K))$ onto Z such that.*

$$f(J_x) = a_f(x), \quad \text{for all } x \in \partial_e K. \tag{7.33}$$

Proof. The mapping $x \rightsquigarrow J_x$ is continuous from the facial topology of $\partial_e K$ to the hull-kernel topology of $\mathrm{Prim}\,A(K)$ (Th. II.6.28). Hence $x \rightsquigarrow f(J_x)$ is continuous in the facial topology of $\partial_e K$ for every $f \in C(\mathrm{Prim}\,A(K))$, and it follows by Theorem II.7.10 that there exists a (unique) $a_f \in Z$ such that (7.33) is valid.

Clearly the mapping $f \rightsquigarrow a_f$ is an isometric isomorphism of $\mathrm{Prim}\,A(K)$ into Z; and it is onto since the mapping $x \rightsquigarrow J_x$ is open as well as continuous in the relevant topologies of $\partial_e K$ and $\mathrm{Prim}\,A(K)$ (Th. II.6.28). □

Notes. Most of the material of this section is quite new.—Theorem II.7.2 was shown to us by Bai Andersen (unpublished), and it has been found independently by Wils [402]. Corollary II.7.5 is due to Alfsen and Andersen [13], and so is Corollary II.7.5 [13].—For simplexes Corollary II.7.5 was proved already by Effros in 1967 [155].—The Theorems II.7.6 and II.7.8 are due to Alfsen and Andersen [13].—The connection between order bounded linear operators and multiplication operators was recognized by Buck in 1961 [93]. In particular he pointed out the non-trivial fact that the algebra of order bounded linear operators is commutative. Theorem II.7.10 appeared in its present form in the paper of Alfsen and Bai Andersen [14], and it was found independently by Wils [402]. Wils was the first to make use of the fact that extreme *contractive* operators are idempotents and determine *splittings* of the positive cone (Prop II.7.13, and Lemma II.8.5. of the next paragraph) [401], [402].—The concept of a "prime simplex" is due to Effros and Kazdan and so is Theorem II.7.15 [161]. They prove that state spaces for harmonic functions are simplexes of the two extreme types. Specifically, one has a *Bauer simplex* if there are no singular boundary points (such as Lebesgue splines), otherwise one has a *prime simplex*. The situation for the heat equation is more complicated allowing also for intermediate cases.—The example given in Proposition II.7.17, is also due to Effros. It was sketched in his 1967 paper on "Structure in simplexes" [155].—Theorem II.7.19 was established by Størmer in 1968 with a somewhat different proof [382].—The Dauns Hofmann Theorem was proved for C^*-algebras by Dauns and Hofmann in 1968, and the generalization to compact convex sets was proved by Alfsen and Bai Andersen in 1969 [14]. (Strictly speaking, the latter result does not generalize the full Dauns Hofmann Theorem, but only the Theorem for C^*-algebras with unit.)

§ 8. Existence and Uniqueness of Maximal Central Measures Representing Points of an Arbitrary Compact Convex Set

Again we shall assume that K is a fixed compact convex set, regularly embedded in $E(\cong A(K)^*)$, and we recall that E is partially ordered with positive cone $\tilde{K}(\cong [A(K)^*]^+)$.

We shall say that two (not necessarily closed) faces F, G of K are *strongly disjoint*, and we shall write $F \between G$, if

$$\operatorname{face}(F \cup G) = F \oplus_c G. \tag{8.1}$$

Note that (8.1) is quite a restrictive definition. It requires that F and G are affinely independent and also that the convex hull of F and G is a face of K.

It is sometimes convenient to avoid the cumbersome constants needed in computations with convex combinations. For this purpose we shall give a simple restatement of the definition of strong disjointness. We recall that for a convex subset F of K we denote by $\tilde{F}$ the cone with base F in E and by $\operatorname{lin} F$ the linear span of F in E.

Proposition II.8.1. *Let F, G be faces of K. Then $F \between G$ iff the following two conditions are both satisfied.*

(i) $\operatorname{lin} F \cap \operatorname{lin} G = \{0\}$.

(ii) $\tilde{F} + \tilde{G}$ *is a hereditary subcone of* $\tilde{K}$.

Proof. By virtue of Proposition II.6.1, F and G are affinely independent iff (i) is valid. It is easily verified that

$$\widetilde{\operatorname{conv}(F \cup G)} = \tilde{F} + \tilde{G},$$

and it follows by Proposition II.2.5 that $\operatorname{conv}(F \cup G)$ is a face of K iff (ii) is valid. The proof is complete. $\square$

We shall say that two points x, y of K are *strongly disjoint*, and we shall write $x \flat y$, if $\operatorname{face}(x) \flat \operatorname{face}(y)$. We shall write $x \not\flat y$ if x, y are not strongly disjoint, and we shall say that a point x of K is *primary* if

$$x = \lambda y + (1-\lambda) z; \quad 0 < \lambda < 1; \quad y, z \in K \quad \Rightarrow \quad y \not\flat z. \tag{8.2}$$

Clearly, every extreme point is primary, ((8.2) is vacuously satisfied); but the reverse implication is inexact in general.

For every point $x \in K$, or more generally for every $x \in \tilde{K}$, we shall denote by V_x the order ideal generated by x in E. Thus

$$V_x = \{ y \in E \mid -\lambda x \leqq y \leqq \lambda x \text{ for some } \lambda \in \mathbb{R}^+ \}. \tag{8.3}$$

It is easily verified that for every $x \in K$:

$$V_x = \operatorname{lin}[\operatorname{face}(x)], \tag{8.4}$$

and

$$V_x^+ = \widetilde{\operatorname{face}(x)}. \tag{8.5}$$

Lemma II.8.2. *Let $y, z \in K$ and let $x = \lambda y + (1-\lambda) z$ where $0 < \lambda < 1$. Then $y \flat z$ iff*

$$V_x = V_y \oplus V_z, \tag{8.6}$$

where this is a direct ordered sum (i.e. the projections onto V_y and V_z are order preserving, cf. [87]).

Proof. To say that V_x is direct ordered sum of V_y and V_z, is equivalent to:

(i)′ $V_y \cap V_z = \{0\}$,

(ii)′ For every $u \in V_x^+$ there exist $v \in V_y^+$, $w \in V_z^+$ such that $u = v + w$.

Using (8.4) and (8.5) we can reduce (i)′ and (ii)′ to the statements (i) and (ii) of Proposition II.8.1 with $F = \operatorname{face}(y)$ and $G = \operatorname{face}(z)$. This completes the proof. $\square$

Proposition II.8.3. *A point* $x \in K$ *is non-primary iff it can be decomposed as a sum* $x = u + v$ *where* $0 \neq u, v$; $u, v \in \tilde{K}$, *and*

$$V_x = V_u \oplus V_v \quad \text{(direct ordered sum).} \tag{8.7}$$

Proof. To say that x is non-primary, means that there exists a decomposition of x as a convex combination of two strongly disjoint elements $y, z \in K$. Writing $u = \lambda y$, $v = (1-\lambda)z$ and using Lemma II.8.2, we convert this into the condition stated in the proposition. □

To fix the ideas, we shall point out what the primary points are in a few simple examples:

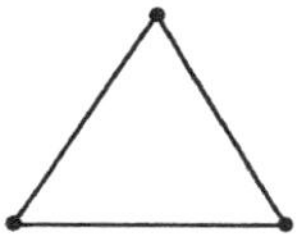

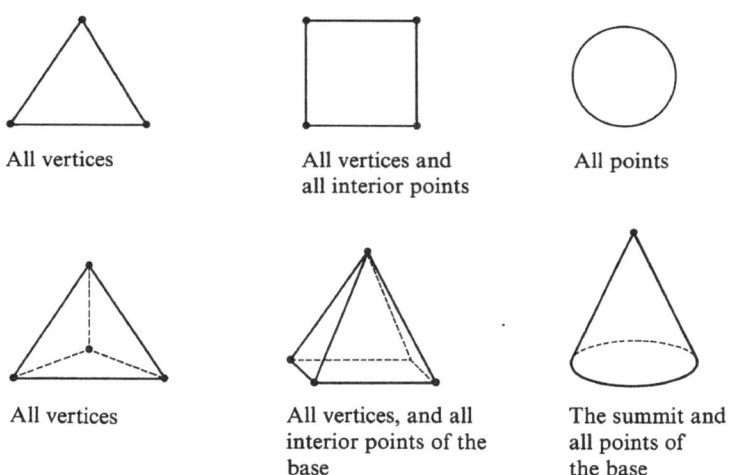

All vertices

All vertices and all interior points

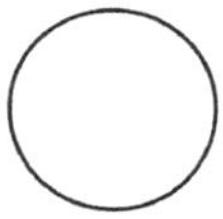

All points

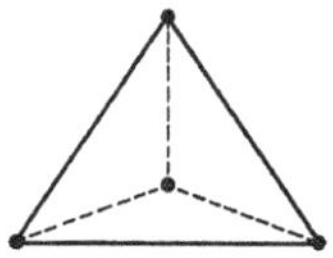

All vertices

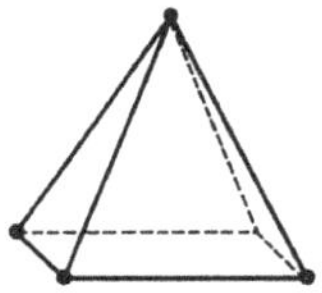

All vertices, and all interior points of the base

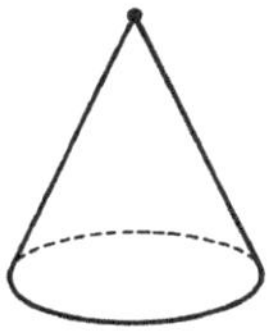

The summit and all points of the base

The spaces V_x will play an important role in the sequel. They are subspaces of the space E in which K is regularly embedded, and we recall that E can be identified with $A(K)^*$ endowed with usual ordering and w^*-topology.

Proposition II.8.4. (V_x, x) *is an order-unit space for every* $x \in K$. *The unit ball* $(V_x)_1$ *is compact in the topology relativized from E, and the space* V_x *itself is complete in order-unit norm.*

Proof. 1. V_x is Archimedean ordered since it is a subspace of $E \cong A(K)^*$, and x is an order-unit by the definition of V_x (i.e. by (8.3)).

2. We have

$$(V_x)_1 = \{y \in E \mid -x \leqq y \leqq x\} = (x - \tilde{K}) \cap (-x + \tilde{K}),$$

and compactness follows from Theorem II.2.6 and Corollary II.2.7.

3. Completeness follows by an elementary argument similar to the proof of Proposition II.1.12. □

For every $x \in K$ we shall denote by Z_x the center of the order-unit space (V_x, x), and we shall call Z_x *the local center at* x.

Lemma II.8.5. *A point $x \in K$ is non-primary iff there exists a linear operator $T: V_x \to V_x$ such that*:

$$T^2 = T, \quad 0 \leqq T \leqq I, \quad T \neq 0, I. \tag{8.8}$$

Proof. 1. Assume first that x is non-primary. By Proposition II.8.3 we have a decomposition

$$V_x = V_y \oplus V_z \quad \text{(direct ordered sum)}, \tag{8.9}$$

where $0 < y, z < x$ and $x = y + z$. This decomposition determines a projection T of V_x onto V_y which is seen to satisfy (8.8).

2. Assume next that T is a linear operator on V_x which satisfies (8.8), and define $y = Tx$ and $z = (I - T)x$. Then it is a routine verification to show that (8.9) is valid. Hence x is non-primary. □

Lemma II.8.6. *Let $x \in K$ be arbitrary. Then there exists a locally convex Hausdorff topology on Z_x such that*

$$\{y \in Z_x \mid 0 \leqq y \leqq x\}$$

is compact.

Proof. By Theorem II.7.10 we can identify Z_x with the space $L_{ob}(V_x)$ of all order bounded linear operators on V_x. By Proposition II.8.4 the unit ball of V_x is compact in the topology $\mathscr{T}$ relativized from E. It follows by a standard argument based on the Tykhonov theorem that

$$\{T \in L_{ob}(V_x) \mid 0 \leqq T \leqq I\}$$

is compact in the topology of pointwise convergence for functions with values in $(V_x, \mathscr{T})$. This completes the proof. □

Theorem II.8.7. *A point $x \in K$ is primary iff the local center at x is trivial.*

Proof. If Z_x is non-trivial, then there exists by Proposition II.7.13 a linear operator T on V_x which satisfies the requirement (8.8) of Lemma II.8.5. Hence x is non-primary.

Conversely, if x is non-primary, then there exists a linear operator T on V_x with the property (8.8) of Lemma II.8.5. Clearly T is order bounded, and T is not a scalar since it is idempotent and different from 0 and I. It follows (by Th. II.7.10) that the local center at x is trivial. The proof is complete. □

We are now going to introduce the most important new concept of this section,—that of a "central measure". In this connection we recall some standard notations: For every measure $\mu \in M(K)$ and every Borel set $B \subset K$ the measure μ_B is the *restriction* of μ to B, i.e. $\mu_B = \chi_B \cdot \mu$ or equivalently $\mu_B(D) = \mu(B \cap D)$ for every Borel set $D \subset K$. For every measure $\mu \in M(K)$ the point $y \in E$ defined by

$$y = \int x \, d\mu(x) \quad \text{(weak integral)}, \tag{8.10}$$

is the *moment* (or "resultant") of μ. If $\mu \in M_1^+(K)$, then the moment of μ is in K, and it is equal to the barycenter of μ (cf. Ch. I. § 2). In the sequel we shall denote the moment of μ by $r(\mu)$.

A measure $\mu \in M_1^+(K)$ is said to be *central* if the barycenters of the probability measures $\lambda^{-1}\mu_B$ and $(1-\lambda)^{-1}\mu_{K \setminus B}$ are strongly disjoint for any Borel set $B \subset K$ such that $\lambda = \mu(B) \in]0,1[$.

To fix the ideas we consider once more the examples on page 173. It is seen that in these particular cases every point $x \in K$ can be represented by a unique central measure which is maximal (among central measures) in Choquet's ordering, and moreover that this measure is concentrated on the set of primary points. We shall see that a similar statement is true in general, and we start by studying a special case where the proof is particularly simple.

In this connection we make the following:

Observation II.8.8 *An order-unit space (A, e) is a vector-lattice iff its center is all of A.*

Proof. The condition is sufficient by Proposition II.7.9, and it is necessary by the definition of center and by the elementary properties of Bauer simplexes (Th. II.4.1–2). □

Theorem II.8.9. *If x is a point of K for which V_x is a vector lattice, then there exists a unique boundary measure $\mu_x \in M_1^+(K)$ representing x, and this measure μ_x is central.*

Proof. 1. To prove the uniqueness it suffices to show that the set $M_x^+(K)'$ of *simple* measures in $M_x^+(K)$ is directed in Choquet's ordering (cf. the proof of Th. II.3.6).

A measure $\mu \in M_x^+(K)'$ is of the form

$$x = \sum_{i=1}^{n} \lambda_i \varepsilon_{x_i},$$

where $x = \sum_{i=1}^{n} \lambda_i x_i$, $\sum_{i=1}^{n} \lambda_i = 1$ and $\lambda_i > 0$ for $i = 1, \dots, n$. In particular $x_i \leqq \lambda_i^{-1} x \in E$ for $i = 1, \dots, n$. Hence $x_1, \dots, x_n \in V_x$. Since V_x is a vector lattice, we can apply the Riesz' decomposition theorem in the form of Proposition II.3.3 to conclude that $M_x^+(K)'$ is directed.

2. We consider an arbitrary Borel set $B \subset K$ such that $\lambda = \mu_x(B) \in]0, 1[$, and we shall prove that the barycenters y, z of the two measures $\lambda^{-1}(\mu_x)_B$, $(1-\lambda)^{-1}(\mu_x)_{K \setminus B}$ are strongly disjoint.

Observe first that V_v is a vector sublattice of V_x for every $v \in V_x$. Thus it follows from the first part of the proof that not only x itself, but also every point $v \in V_x \cap K$, can be represented as the barycenter of a unique boundary measure $\mu_v \in M_1^+(K)$. This result can be extended by linearity to all of $V_x = \mathrm{lin}(V_x \cap K)$, and we conclude that every point $v \in V_x$ is the moment of a unique boundary measure $\mu_v \in M_{\mathbb{R}}(K)$.

For every $v \in V_x$ we now define

$$T v = r((\mu_v)_B). \tag{8.11}$$

Clearly $0 \leqq T u \leqq T v$ whenever $0 \leqq u \leqq v$. Hence T is a linear map of V_x into itself, and $0 \leqq T \leqq I$.

By virtue of (8.11) we have $\mu_{Tv} = (\mu_v)_B$ for every $v \in V_x$. This implies

$$T^2 v = T(T v) = r((\mu_{Tv})_B) = r((\mu_v)_B) = T v$$

for every $v \in V_x$. Hence $T^2 = T$.

By definition $y = \lambda^{-1} T x$ and $z = (1-\lambda)^{-1}(I - T)x$, and it is a routine verification to show that

$$V_x = V_y \oplus V_z \quad \text{(direct ordered sum)}.$$

Now it follows from Lemma II.8.2 that $y \curlywedge z$, and the proof is complete. □

The first statement of Theorem II.8.9 is a local version of Choquet's uniqueness theorem. Note also that the central measure μ_x above will be maximal among *all* measures in $M_x^+(K)$, and not only among the central ones. Such a central measure will of course not exist in general.

In the following we shall reduce the general case to the special situation of Theorem II.8.9 by a "lifting" technique. This technique can be explained either in terms of tensor products of partially ordered vector spaces, or by application of appropriate spaces of bilinear forms. To avoid the introduction of new machinery, we shall prefer the latter approach.

We shall assume that x is an arbitrary, but fixed, point of K, and we consider the vector space of all bilinear forms on $V_x \times A(K)$ (partially) ordered by the cone B_x^+ of all forms b such that:

$$b(y,a) \geqq 0, \quad \text{all } y \geqq 0, \quad a \geqq 0. \tag{8.12}$$

Observe that B_x^+ has a w^*-compact base C_x defined by evaluation at the pair $(x, 1_K)$ of the two order-units. Thus:

$$C_x = \{b \in B_x^+ \mid b(x, 1_K) = 1\}. \tag{8.13}$$

We define $B_x = B_x^+ - B_x^+$ and observe that (B_x, C_x) is a base-norm space.

By assumption K is regularly embedded in E, hence $E \cong A(K)^*$. Specifically, to every point $y \in E$ corresponds the linear functional $a \rightsquigarrow \bar{a}(y)$ (cf. Ch.II, § 2).

Now we obtain a natural ("restriction") map $\pi: B_x \to A(K)^* \cong E$, by fixing the first variable at the order-unit x. Specifically:

$$\bar{a}(\pi b) = b(x,a), \quad \text{for all } b \in B_x, \quad a \in A(K). \tag{8.14}$$

It is easily verified that π is order preserving and maps C_x into K.

We denote by b_x the bilinear form on $V_x \times A(K)$ defined by

$$b_x(y,a) = \bar{a}(y); \tag{8.15}$$

and we observe that $\pi b_x = x$, since

$$\bar{a}(\pi b_x) = b_x(x,a) = \bar{a}(x),$$

for all $a \in A(K)$.

Since the two spaces V_x and $A(K)$ are already in duality (under the pairing $(y,a) \rightsquigarrow \bar{a}(y)$), every $b \in B_x$ will define a linear map from the one space into the dual of the other. We shall be particularly interested in the operator $R_b: V_x \to A(K)^* \cong E$, which is hereby associated with an element b of B_x. Specifically:

$$\bar{a}(R_b y) = b(y,a) \tag{8.16}$$

for all $y \in V_x$, $a \in A(K)$.

Observe that

$$R_b x = \pi b, \tag{8.17}$$

since

$$\bar{a}(R_b x) = b(x,a) = \bar{a}(\pi b),$$

for all $a \in A(K)$.

Note also that R_{b_x} is the identity operator of V_x onto itself, i.e.

$$R_{b_x} y = y, \quad \text{for all } y \in V_x, \tag{8.18}$$

since

$$\bar{a}(R_{b_x} y) = b_x(y,a) = \bar{a}(y),$$

for all $y \in V_x$, $a \in A(K)$.

We provide $\mathrm{Lin}(V_x, E)$ with the natural ordering. Thus, an operator $R: V_x \to E$ is said to be *positive* if $Ry \in E^+$ for every $y \in V_x^+$. Then we shall have

$$b \in B_x^+ \Rightarrow R_b \geqq 0, \tag{8.19}$$

since $b \geqq 0$ and $y \geqq 0$ implies

$$\bar{a}(R_b y) = b(y,a) \geqq 0$$

for all $a \in A(K)$.

For every $b \in B_x^+$ we shall denote by V_b the order ideal generated by b in B_x. Thus

$$V_b = \{d \in B_x \mid -\lambda b \leqq d \leqq \lambda b \text{ for some } \lambda \in \mathbb{R}^+\}. \tag{8.20}$$

Clearly, (V_b, b) is an order-unit space for every $b \in B_x^+$.

By (8.18) and (8.19) we shall have the implication:

$$b \in B_x, \quad 0 \leqq b \leqq b_x \Rightarrow 0 \leqq R_b \leqq I, \tag{8.21}$$

where I is the identity operator of V_x onto itself.

Upon this fact rests the following simple, but useful:

Observation II.8.10. *For given $x \in K$ and $b \in V_{b_x}$, the operator R_b maps V_x into itself. In fact, R_b is an order bounded linear operator on (V_x, x).*

Proposition II.8.11. *If x is a given point of K, then the ("restriction") map $\pi: B_x \to E$ maps V_{b_x} into the center of (V_x, x).*

Proof. Let $b \in V_{b_x}$. By Observation II.8.10 above $R_b \in L_{ob}(V_x)$, and by (8.17) $R_b x = \pi b$. Hence we obtain the element πb of V_x by acting on the order unit x with an order bounded linear operator. Now it follows by Theorem II.7.10 (and by the definition (7.13)) that πb is in the center of (V_x, x). The proof is complete. ▯

Proposition II.8.11 provides the first half of the correspondence between V_{b_x} and the center Z_x of (V_x, x). The second half is less obvious, since there is no natural "lifting" of elements in V_x to bilinear forms. However, it is not necessary to lift arbitrary elements of V_x, but only the *central* ones. These correspond to order bounded linear operators, and they can be lifted. In fact, a linear operator on V_x acts by operator multiplication on the particular operators R_b, and hence they can be made to act directly on the elements $b \in V_{b_x}$.

Specifically, we associate with every $T \in L_{ob}(V_x)$ a "lifted" operator $\tilde{T}$ between bilinear forms on $V_x \times A(K)$, defined by:

$$[\tilde{T} b](y,a) = \bar{a}(T R_b y), \quad \text{for } y \in V_x, \quad a \in A(K). \tag{8.22}$$

Proposition II.8.12. *If x is a given element of K and $T \in L_{ob}(V_x)$, then the operator $\tilde{T}$ defined by* (8.22) *maps V_{b_x} into itself. In fact, $\tilde{T}$ is an order bounded linear operator on (V_{b_x}, b_x).*

Proof. We shall be through if we can establish the implication:

$$b \in B_x^+, \quad 0 \leqq T \leqq I \Rightarrow 0 \leqq \tilde{T} b \leqq b. \tag{8.23}$$

Let $b \in B_x^+$ and $0 \leqq T \leqq I$. By (8.19) $R_b \geqq 0$, i.e. $R_b y \geqq 0$ for all $y \in V_x^+$. It follows that

$$0 \leqq T R_b y \leqq R_b y$$

for all $y \in V_x^+$, and hence

$$0 \leqq \bar{a}(T R_b y) \leqq \bar{a}(R_b y)$$

for all $y \in V_x^+, a \in A(K)^+$.

Using the definitions (8.22) and (8.16), we obtain

$$0 \leqq [\tilde{T} b](y,a) \leqq b(y,a)$$

for all $y \in V_x^+$, $a \in A(K)^+$; and (8.23) is proved. □

It should be noted that an order bounded linear operator T on V_x is completely determined by its lifting $\tilde{T}$ to V_{b_x}. In fact it follows by (8.18) and (8.22) that

$$\bar{a}(T y) = \bar{a}(T R_{b_x} y) = [\tilde{T} b_x](y,a), \tag{8.24}$$

for all $y \in V_x, a \in A(K)$.

Theorem II.8.13. *If x is any given point of K, then V_{b_x} is a vector lattice, and π is an isomorphism of the order-unit space (V_{b_x}, b_x) onto the order-unit space (Z_x, x), i.e. onto the local center at the point x.*

Proof. 1. We shall first prove that $\pi: V_{b_x} \to Z_x$ is an isomorphism by constructing a two-sided inverse $\sigma: Z_x \to V_{b_x}$.

It follows from Theorem II.7.10 that each central element z of (V_x, x) corresponds to an order-bounded linear operator T_z on V_x, which maps the order-unit x into the given central element z. In symbols:

$$z = T_z x, \quad \text{for all } z \in Z_x . \tag{8.25}$$

We repeat the same process in V_{b_x} with $\tilde{T}_z$ in the place of T_z. Thus we define:

$$\sigma(z) = \tilde{T}_z b_x, \quad \text{for all } z \in Z_x . \tag{8.26}$$

By the definitions (8.14), (8.26) of π, σ and by the formulas (8.24), (8.25), we obtain for all $z \in Z_x$, $a \in A(K)$:

$$\begin{aligned} \bar{a}(\pi(\sigma z)) &= \bar{a}(\pi(\tilde{T}_z b_x)) \\ &= [\tilde{T}_z b_x](x, a) \\ &= \bar{a}(T_z x) \\ &= \bar{a}(z) . \end{aligned}$$

Hence:

$$[\pi \circ \sigma](z) = z, \quad \text{for all } z \in Z_x . \tag{8.27}$$

It follows from Theorem II.7.10 that order bounded linear operators on order-unit spaces are completely determined by their values at the order-unit. By (8.17) and (8.25), we obtain for an arbitrary $b \in V_{b_x}$:

$$R_b x = \pi b = T_{\pi b} x .$$

Hence we obtain the general formula

$$T_{\pi b} = R_b, \quad \text{for all } b \in V_{b_x} . \tag{8.28}$$

By the definition (8.26) of σ, by the formulas (8.24), (8.28) and by the definition (8.17) of R_b, we obtain for all $b \in V_{b_x}$, $y \in V_x$, $a \in A(K)$:

$$\begin{aligned} [\sigma(\pi b)](y, a) &= [\tilde{T}_{\pi b} b_x](y, a) \\ &= \bar{a}(T_{\pi b}\, y) \\ &= \bar{a}(R_b\, y) \\ &= b(y, a) . \end{aligned}$$

Hence:

$$[\sigma \circ \pi](b) = b, \quad \text{for all } b \in V_{b_x} . \tag{8.29}$$

It follows from (8.27) and (8.29) that π is a linear isomorphism of V_{b_x} onto Z_x with inverse σ.

We have noted before that π is an order preserving map, and that $\pi b_x = x$. If $z \in Z_x^+$, then T_z is a positive linear operator on V_x (Th. II.7.10), and it follows by (8.23) that $\tilde{T}_z$ is a positive linear operator on V_{b_x}; and so $\sigma z = \tilde{T}_z b_x \geqq 0$. Hence σ is order preserving, and π is an isomorphism of the order-unit space (V_{b_x}, b_x) onto the order-unit space (Z_x, x).

2. The local center Z_x is a vector lattice by Proposition II.7.9, and it follows by the first part of the proof that V_{b_x} is a vector lattice as well. The proof is complete. ▯

Corollary II.8.14. *If x is a given point of K and b is a primary point of C_x, then πb is a primary point of K.*

Proof. Let V_b be the order ideal generated by b in V_{b_x}, let $z = \pi b$ and let V_z be the order ideal generated by z in V_x. We observe that V_b is a vector sublattice of V_{b_x}, hence the order-unit space (V_b, b) is equal to its center (Obs. II.8.8), and this center is trivial since b is primary (Th. II.8.7). Thus we have the following formula (by which $b \in \partial_e C_x$):

$$V_b = \{\alpha b \mid \alpha \in \mathbb{R}\}. \tag{8.30}$$

Let y be an arbitrary central element of V_z. Then there exists an $S_y \in L_{ob}(V_z)$ such that $S_y z = y$, and there exists also a $T_z \in L_{ob}(V_x)$ such that $T_z x = z$ and $\mathrm{Range}(T_z) = V_z$ (Th. II.7.10). The operator $T = S_y T_z$ is well defined on V_x, and $Tx = y$. Hence it follows (again by Th. II.7.10) that $y \in Z_x$.

Now it follows from Theorem II.8.13 that y is the image by π of a unique element of V_{b_x}, and this element must be in V_b since $y \in V_z$. Hence we have $y = \pi(\alpha b) = \alpha z$ for some $\alpha \in \mathbb{R}$.

We have proved that the center of (V_z, z) is trivial. By Theorem II.8.7, z must be a primary point of K, and we are through. ▯

We are now going to establish the existence and uniqueness theorem for maximal central measures, and we shall first prove some auxiliary results. The first of these is actually a general property of moments of transported measures. For the sake of brevity, we shall state it in the special setting of the present chapter.

Recall that a measure ν on C_x is *transported* to a measure $\pi\nu$ on K defined by

$$[\pi \nu](B) = \nu(\pi^{-1}(B)) \tag{8.31}$$

for all Borel sets $B \subset K$; and recall also that

$$\int_B f \, d(\pi\nu) = \int_{\pi^{-1}(B)} f \circ \pi \, d\nu \tag{8.32}$$

for all Borel sets $B \subset K$ and all $f \in C_{\mathbb{R}}(K)$.

Lemma II.8.15. *If ν is a measure on C_x and B is Borel subset of K, then*

$$\pi(r(\nu_{\pi^{-1}(B)}))=r((\pi\nu)_B). \tag{8.33}$$

Proof. For every $a\in A(K)$ we obtain by (8.32) and the definition of moments:

$$\begin{aligned}\bar{a}(\pi(r(\nu_{\pi^{-1}(B)})))&=[\bar{a}\circ\pi](r(\nu_{\pi^{-1}(B)}))\\&=\int_{\pi^{-1}(B)}\bar{a}\circ\pi\, d\nu\\&=\int_B \bar{a}\, d(\pi\nu)\\&=\bar{a}(r((\pi\nu)_B))\end{aligned}$$

This completes the proof. □

The next lemma is an adaption of Proposition I.2.3 to the case of central measures. First we state the following:

Sublemma II.8.16. *Let $\mu\in M_1^+(K)$ be a central measure and let $\{B_1,\dots,B_n\}$ be a Borel partition of K such that $\lambda_i=\mu(B_i)\neq 0$ for $i=1,\dots,n$. If x_i is the barycenter of $\lambda_i^{-1}\mu_{B_i}$ for $i=1,\dots,n$, then the measure*

$$\nu=\sum_{i=1}^{n}\lambda_i\varepsilon_{x_i} \tag{8.34}$$

is central.

Proof. Let B be an arbitrary Borel subset of K such that $\lambda=\nu(B)\in]0,1[$, and let y,z be the barycenters of $\lambda^{-1}\nu_B$ and $(1-\lambda)^{-1}\nu_{K\setminus B}$, respectively.

Now define $J=\{i\mid x_i\in B,\ i=1,\dots,n\}$ and $D=\bigcup_{i\in J}B_i$. Clearly $\lambda=\mu(D)$, and

$$y=\lambda^{-1}r(\nu_B)=\lambda^{-1}\sum_{j\in J}\lambda_i x_i=\lambda^{-1}r(\mu_D).$$

Hence, y is the barycenter of $\lambda^{-1}\mu_D$. Similarly we observe that z is the barycenter of $(1-\lambda)^{-1}\mu_{K\setminus D}$. Since μ is supposed to be central, we shall have $y\,\delta\, z$, and the proof is complete. □

Lemma II.8.17. *Every central measure $\mu\in M_x^+(K)$ can be vaguely approximated by central simple measures in $M_x^+(K)$.*

Proof. The construction of the approximating measures is identical with that of the proof of Proposition I.2.3, and the approximating measures are all central by virtue of the sublemma above. □

The following elementary fact on centers of order-unit spaces is implicite in the general theory of the preceding paragraph, but we state it as a proposition for later references.

Proposition II.8.18. *Let (A,e) be an order-unit space, let $a\in Z(A)$, and let T_a be the order bounded linear operator on A which corresponds to a. Then the following two statements are equivalent:*

(i) $$T_a^2 = T_a$$

(ii) $$a\wedge(e-a)=0 \quad \text{(g.l.b. in } Z(A)).$$

Proof. By Proposition II.7.9 and Theorem II.7.10, $Z(A)$ is an M-space and a commutative Banach algebra where the multiplication is given by $ab=T_a b$. Moreover, $Z(A)$ can be identified with the function space C_f (provided with pointwise lattice operations and multiplication). If $g\in C_f$, then

$$g^2=g \;\Leftrightarrow\; \text{Range}(g)\subset\{0,1\} \;\Leftrightarrow\; g\wedge(1-g)=0,$$

and the equivalence of (i) and (ii) follows. □

An element a of an order-unit space (A,e) is said to be a *central idempotent* if it is in $Z(A)$ and satisfies the two equivalent requirements (i), (ii) of Proposition II.8.18.

Remark. The central idempotents of an order-unit space (A,e) form a Boolean algebra with $a\wedge b=T_a b=T_b a$, $a\vee b=a+b-a\wedge b$ and $a^c=e-a$. This Boolean algebra is complete if $\{a\in Z(A)\,|\,0\leqq a\leqq e\}$ is compact in some locally convex Hausdorff topology on $Z(A)$ (as e.g. for the order-unit spaces (V_x,x) defined in this paragraph, cf. Lem. II.8.6). The proof of the first statement is straightforward. The proof of the second invokes the Krein-Milman theorem as in the proof of Proposition II.7.13. We shall not need these results in the sequel.

Our next lemma is the last one before the main theorem. It is essentially a restatement of an argument which has already been used in the proof of Theorem II.8.9.

Lemma II.8.19. *Let x be a given point of K. Then a measure $\mu\in M_x^+(K)$ is central iff $r(\mu_B)$ is a central idempotent of (V_x,x) for every Borel subset B of K.*

Proof. 1. We first assume the condition of the lemma, and we shall show that μ must be a central measure.

To this end we consider an arbitrary Borel set $B \subset K$ such that $\lambda = \mu(B) \in]0,1[$, and we define y, z to be the barycenters of the measures $\lambda^{-1}\mu_B$ and $(1-\lambda)^{-1}\mu_{K \setminus B}$, respectively.

By assumption $r(\mu_B)$ is a central idempotent of (V_x, x), and we denote the corresponding order bounded linear operator on V_x by T. Then we shall have

$$0 \leqq T \leqq I, \qquad T^2 = T, \qquad Tx = r(\mu_B). \tag{8.35}$$

It follows from the last equality of (8.35) that

$$y = \lambda^{-1} Tx, \qquad z = (1-\lambda)^{-1}(I-T)x, \tag{8.36}$$

and it is seen by an easy verification based on the order properties and idempotence that

$$V_x = V_y \oplus V_z \quad \text{(direct ordered sum)}. \tag{8.37}$$

Now it follows from Lemma II.8.2 that $y \between z$. Hence μ is a central measure.

2. Next we assume that μ is central, and we consider an arbitrary Borel set $B \subset K$. If $\mu(B) = 0, 1$ then $r(\mu_B) = 0, x$; hence $r(\mu_B)$ is a central idempotent of (V_x, x) in these two cases. If $\lambda = \mu(B) \in]0,1[$, then we define y, z to be the barycenters of $\lambda^{-1}\mu_B$ and $(1-\lambda)^{-1}\mu_{K \setminus B}$ respectively, and we shall have $y \between z$ since μ is assumed to be central. By Lemma II.8.2 we shall have a decomposition of the type (8.37) also in this case, and we now define T to be the corresponding projection of V_x onto V_y.

Clearly $0 \leqq T \leqq I$ and $T^2 = T$. Moreover, we can decompose

$$x = \lambda y + (1-\lambda) z,$$

and it follows that $Tx = \lambda y = r(\mu_B)$. Hence $r(\mu_B)$ is a central idepotent of (V_x, x), associated with the idempotent order bounded linear operator T on V_x. The proof is complete. □

A measure $\mu \in M_x^+(K)$ will be called a *maximal central measure* if it is central and if it is maximal in Choquet's ordering among all central measures in $M_x^+(K)$.

Recall that a measure $\mu \in M_1^+(K)$ is said to be *pseudo carried* by a subset D of K if $\mu(C) = 0$ for every compact G_δ-set $C \subset K$ such that $C \cap D = 0$. (This is equivalent to $\mu_0^*(D) = 1$, where μ_0^* is the outer Baire measure associated with μ.)

Theorem II.8.20. *(Wils) A point x of a compact convex set K can be represented as the barycenter of a unique maximal central measure μ, and μ is pseudo-carried by the set of primary points of K.*

Proof. 1. It follows from Theorem II.8.9 that the point b_x of C_x is represented by a unique measure $\tilde{\mu}$ which is maximal in $M_1^+(C_x)$, and that this measure is central. We shall show that the measure $\mu=\pi\tilde{\mu}$ has all the desired properties.

Clearly $\mu\in M_1^+(K)$, and it follows by application of Lemma II.8.15 with $B=K$ that $x=\pi(b_x)=r(\mu)$ so that μ has barycenter x.

To prove that μ is central, we shall apply the criterion of Lemma II.8.19. Thus we consider an arbitrary Borel set $B\subset K$, and by Lemma II.8.15

$$r(\mu_B)=\pi(r(\tilde{\mu}_{\pi^{-1}(B)})). \tag{8.38}$$

Now $r(\tilde{\mu}_{\pi^{-1}(B)})$ must be a central idempotent of (V_{b_x}, b_x) since $\tilde{\mu}$ is a central measure on C_x (Lem. II.8.19). The concept of a central idempotent is defined in terms of the order properties of the center (cf. statement (ii) of Prop. II.8.18); hence it follows by Theorem II.8.13 that an element b of V_{b_x} is a central idempotent iff πb is a central idempotent of (V_x, x). Applying this to (8.38) we conclude that $r(\mu_B)$ is a central idempotent of (V_x, x). This completes the verification that μ is a central measure on K.

Proceding as in the proof of Theorem II.3.6 with application of Lemma II.8.17 rather than of Proposition I.2.3, we can reduce the proof that μ is a maximal central measure to the verification that $\nu\prec\mu$ for every central measure

$$\nu=\sum_{i=1}^{n}\lambda_i\varepsilon_{x_i}, \tag{8.39}$$

where $x=\sum_{i=1}^{n}\lambda_i x_i$, $\sum_{i=1}^{n}\lambda_i=1$ and $\lambda_i>0$ for $i=1,\dots,n$.

Since ν is a central measure, the points $\lambda_1 x_1,\dots,\lambda_n x_n$ are central idempotents of (V_x, x) (Lem. II.8.19). By Theorem II.8.13 there exist points $b_1,\dots,b_n\in V_{b_x}$ such that

$$b_x=\sum_{i=1}^{n}\lambda_i b_i, \qquad \pi b_i=x_i \quad \text{for } i=1,\dots,n.$$

It is easily verified that $b_i\in C_x$ for $i=1,\dots,n$, and hence we can define a measure $\tilde{\nu}$ on C_x by writing

$$\tilde{\nu}=\sum_{i=1}^{n}\lambda_i\varepsilon_{b_i}.$$

Clearly $\tilde{\nu}$ has barycenter b_x and $\pi\tilde{\nu}=\nu$. By the maximality and uniqueness of $\tilde{\mu}$, we shall have $\tilde{\nu}\prec\tilde{\mu}$. Since Choquet's ordering is preserved under continuous affine maps, this entails

$$\nu=\pi\tilde{\nu}\prec\pi\tilde{\mu}=\mu\,.$$

Hence we have proved that μ is a maximal central measure representing x.

2. To prove that μ is pseudo carried by the set of primary points of K, we consider a compact G_δ-set $D\subset K$ which does not contain any primary points. Now $\pi^{-1}(D)$ is a closed, and hence compact, G_δ-set in C_x.

Also we must have

$$\pi^{-1}(D)\cap\partial_e C_x=\emptyset, \tag{8.40}$$

for if a point $b\in\pi^{-1}(D)$ was extreme in C_x, then it would be primary, and by Corollary II.8.14 πb should be a primary point of K contained in D, contrary to assumption.

Now it follows from (8.40) and from Bishop-de Leeuw's theorem (Cor. I.4.12) that

$$\mu(D)=\tilde{\mu}(\pi^{-1}(D))=0\,;$$

and the proof is complete. □

Remark. At the end of Ch. II § 6 we made some comments on the connections with C^*-algebras, and we feel that some additional comments should be made at this point: If p is a state on a C^*-algebra $\mathscr{A}$ and π_p is the corresponding cyclic representation, then the order ideal $V_p\subset\mathscr{A}^*$ can be identified with the commutant $\pi_p(\mathscr{A})'$ of $\mathscr{A}$ in the representation π_p, and the local center Z_p can be identified with the center $\pi_p(\mathscr{A})''\cap\pi_p(\mathscr{A})'$ of the weak closure of $\mathscr{A}$ in the representation π_p. If p and q are states on $\mathscr{A}$, then the relation $p\,\delta\,q$ defined above, is equivalent with the corresponding relation in C^*-theory, i.e. π_p and π_q shall not admit any two mutually unitarily equivalent subrepresentations. Moreover, the concept of a "maximal central measure" will be equivalent to Sakai's notion of a central measure for C^*-algebras [357]. Note also that the lifting technique of the present paragraph corresponds to a reduction to the "multiplicity free" case, and that the space B_x is used in much the same way as the algebra generated by $\pi_p(\mathscr{A})\cup\pi_p(\mathscr{A})'$.

Notes. In 1965 Sakai proved the existence and uniqueness of a maximal measure which represents a given state of a separable C^*-algebra and is concentrated on the primary states (factor states) [357].—This theorem was generalized to arbitrary C^*-algebras and proved by means of the Bishop-de Leeuw theorem by Wils in 1968 [400].—Finally, Wils stated and proved it for arbitrary compact convex sets later in the same year [401].—The material of the present paragraph is entirely due to Wils, although the details of proof may differ slightly from his; and the reader is referred to Wils' papers [400], [401], [402] for further information on the subject.

Appendix

Some Recent Developments

Many important contributions to the theory of compact convex sets have been made after the manuscript to the present book was completed, and it is impossible here to give a complete survey,—not even of those works which are directly connected with the material treated in the book. Therefore we shall content ourselves with a brief account which may serve as a guide to recent literature.

1. It has been pointed out by Mokobodzki and Rogalski [346] that Theorem II.2.6 can be improved. In fact, *every affine function of first Baire class on K is pointwise limit of a bounded sequence from $A(K)$*. This result can be most easily obtained by application og Theorem II.2.6 together with a general result on sequential limits in Banach spaces proved by McWilliams in 1964 [431].

2. Theorem I.3.8 was proved under the assumption of metrizability, which was needed for the application of the *disintegration theorem* of [89]. For a more general treatment of disintegration one may consult Ionesco Tulcea's recent monograph on lifting of of measures [426].

3. It is possible to give a proof of the Choquet-Bishop-de Leeuw Theorem which at once produces a *simplicial* boundary measure representing a given point of K. This has been done by F.G. Vincent-Smith [394], and independently by Andenæs who derived it from a general theorem on the existence of Hahn-Banach extensions which are maximal on a given cone [20].

4. Lazar and Lindenstrauss have recently given a matrix representation of the preduals of L^1-spaces [266]. Specializing to matrices with positive elements and dualizing, one gets a representation of separable Choquet simplexes. More specifically, to every separable Choquet simplex there corresponds a (non-unique) infinite matrix $\{\lambda_{ij}\}$ such that $\lambda_{ij}>0$ for all i,j, $\lambda_{ij}=0$ when $i<j$, and $\lambda_{i1}+\cdots+\lambda_{ii}=1$ for all i. This matrix determines K completely. In fact, K is the *projective limit* of the sequence

$$\Delta_1 \xleftarrow{\varphi_1} \Delta_2 \xleftarrow{\varphi_2} \Delta_3 \xleftarrow{\varphi_3} \cdots$$

of "collapsing" n-simplexes Δ_n, where φ_n maps the n-simplex Δ_n onto its "base" Δ_{n-1} in such a way that the first n vertices are fixed and the $(n+1)$-th vertex goes to the point with barycentric coordinates $\{\lambda_{n1}, \ldots, \lambda_{nn}\}$.

5. The notion of facial topology has been generalized and axiomatized by Rogalski [347] [348]. A face F of K is said to be *parallelizable* if every $x \in K$ admits a decomposition $x = \lambda y + (1-\lambda) z$ with $v \in F$, $z \in F'$ and with a *unique* coefficient λ. A topology on $\partial_e K$ is said to be *facial* if the closed sets are of the form $F \cap \partial_e K$ with $F \in \mathscr{F}$ for some collection $\mathscr{F}$ of closed parallelizable faces closed under under finite convex combinations and arbitrary intersections. The reader is referred to the above papers for further information on general facial topologies and their relation to the particular facial topology studied in this book.

6. T. Bai Andersen [22] has proved the following theorem on dominated extensions of continuous affine functions on split faces, which is an improvement of Theorem II.6.15: *Let F be a closed split face of K and g a continuous function on K and $b \in A(K)$ and $a_0 \in A(F)$, and assume that $b < g$ and $b \mid F \leq a_0 \leq g \mid F$. Then there exists an $a \in A(K)$ such that $b \leq a \leq g$ and $a \mid F = a_0$.* Note that the ε in Theorem II.6.15 is eliminated; and that strict inequality is retained only in the condition $b < g$. This is achieved by an inductive argument which is a modification of a construction devised by Pelcynski for the study of simultaneous extensions in $C_{\mathbb{R}}(X)$. It follows from Bai Andersen's theorem that not only is a split face *Archimedean* with characteristic number equal to 1, but extensions can be chosen with *strict* preservation of norm, i.e. one can take $\varepsilon = 0$ in (5.17). (For simplexes this is a consequence of Theorem II.5.9.)

7. The generalization of Choquet boundary theory to complex function spaces is straightforward up to a certain point. When estimation of *norms* are required, one can no longer rely on the decomposition into real and imaginary parts; and when *complex orthogonal measures occur*, one will run into more severe difficulties which in interesting applications are related to *analyticity*. O. Hustad has recently proved a *norm preserving* Choquet theorem in the complex case [425], and Asimow has used geometric arguments to study *peak sets* of complex funditon spaces [29]. The complexification of Bai Andersen's theorem (cf. point 6 above) has been performed by Alfsen and Hirsberg [15]. Combining it with Theorem II.6.12 one obtains a generalized verson of Bishop's Rudin-Carleson Theorem (cf. [15]). Some closely related work have been done by Björk [407] and Briem [408]. Björk's result is slightly less general and avoids Bai Andersen's construction. Briem's work involves Bai Andersen's construction, but it avoids the geometry of the state space which

is essential in the Alfsen-Hirsberg paper. Instead, Briem makes use of a recent result of Rao on the existence of *measurable selections* of representing boundary measures which is of considerable interest in itself [432].

References

1. Aarnes, J. F., Effros, E. G., Nielsen, O. A.: Locally compact spaces and two classes of C^*-algebras. Pacific J. Math. **34** (1970) (to appear).
2. Ajlani, M., Goullet de Rugy, A.: Les cônes biréticulés. C. R. Acad. Sci. Paris **270,** 242—245 (1970).
3. Akemann, C. A., Russo, B.: Geometry of the unit sphere of a C^*-algebra and its dual. Pacific J. Math. **32,** 575—585 (1970).
4. Alaoglu, L.: Weak topologies of normed linear spaces. Ann. of Math. (2) **41,** 252—267 (1940).
5. Alfsen, E. M.: On the geometry of Choquet simplexes. Math. Scand. **15,** 97—110 (1964).
6. — A measure theoretic characterization of Choquet simplexes. Math. Scand. **17,** 106—112 (1965).
7. — On the decomposition of a Choquet simplex into a direct convex sum of complementary faces. Math. Scand. **17,** 169—176 (1965).
8. — Boundary values for homomorphisms of compact convex sets. Math. Scand. **19,** 113—121 (1966).
9. — Borel structure on a metrizable Choquet simplex and on its extreme boundary. Math. Scand. **19,** 161—171 (1966).
10. — On Choquet simplexes. Proc. Coll. on Convexity, Copenhagen 1965, 1—8 (1967).
11. — On the Dirichlet problem of the Choquet boundary. Acta Math. **120,** 149—159 (1968).
12. — Facial structure of compact convex sets. Proc. London Math. Soc. **18,** 385—404 (1968).
13. — Bai Andersen, T.: Split faces of compact convex sets. Aarhus University Preprint Series 1968/69 No. 32 (1969).
14. — — On the concept of center for $A(K)$-spaces (to appear).
15. — Hirsberg, B.: On dominated extensions in linear subspaces of $C_{\mathbb{C}}(X)$. Preprint Series, Oslo University 1970 No. 1.
16. — Nordseth, T.: Vertices of Choquet simplexes. Math. Scand. **23,** 171—176 (1968).
17. — Skau, Chr.: Existence et unicité des représentations intégrales par les mesures simpliciales sur la frontière extrémale d'un convex compact. C. R. Acad. Sci. Paris **268,** 1390—1393 (1969).
18. — — Simplicial decomposition of boundary measures on convex compact sets. Math. Scand. **26,** 62—72 (1970).
19. Amir, D., Lindenstrauss, J.: The structure of weakly compact sets in Banach spaces. Ann. of Math. **88,** 35—46 (1968).
20. Andenæs, P. R.: Hahn-Banach extensions which are maximal on a given cone. Math. Annalen **188,** 90—96 (1970).
21. Andersen, T. Bai: On multipliers and order bounded operators in C^*-algebras Proc. Amer. Math. Soc. **25,** 869—899 (1970).

22. Andersen, T. Bai: Extension of continuous affine functions on split faces (to appear in Math. Scand.).
23. — (cf. also Alfsen, E. M., Andersen, T. Bai).
24. Arens, R. A., Singer, I. M.: Function values as boundary integrals. Proc. Amer. Math. Soc. **5**, 735—745 (1954).
25. Andô, T.: On fundamental properties of a Banach space with a cone. Pacific J. Math. **12**, 1163—1169 (1962).
26. Asimov, L.: Well-capped convex cones. Pacific J. Math. **26**, 421—431 (1968).
27. — Directed Banach spaces of affine functions. Trans. Amer. Math. Soc. **143**, 117—132 (1969).
28. — Extremal structure of well-capped convex sets. Trans. Amer. Math. Soc. **138**, 363—375 (1969).
29. — Decomposable compact convex sets and peak sets for function spaces. Proc. Amer. Math. Soc. **25**, 75—79 (1970).
30. — Extensions of continuous affine functions (to appear).
31. — Ellis, A. J.: Facial decomposition of linearly compact simplexes and separation of functions on cones (to appear).
32. Asplund, E.: Frêchet differentiability of convex functions. Acta Math. **121**, 31—47 (1968).
33. Aumann, G.: Erweiterungen von additiven monotonen Funktionen. Arch. Math. **8**, 422—427 (1957).
34. Banach, S.: Théorie des operations linéaires. Warszawa 1932.
35. Bastiani, A.: Cônes convexes et pyramides convexes. Ann. Inst. Fourier, **9**, 249—292 (1959).
36. Bauer, H.: Geordnete Gruppen mit Zerlegungseigenschaft. Bayer. Akad. der Wissenschaften, München 1958, 25—36.
37. — Un problème de Dirichlet pour la frontière de Silov d'un espace compact. C. R. Acad. Sci., Paris **247**, 843—846 (1958).
38. — Minimalstellen von Funktionen und Extremalpunkte. Arch. Math. **9**, 389—393 (1958).
39. — Minimalstellen von Funktionen und Extremalpunkte II. Arch. Math. **11**, 200—205 (1960).
40. — Schilowscher Rand und Dirichletsches Problem. Ann. Inst. Fourier **11**, 89—136 (1961).
41. — Konvexität in topologischen Vektorräumen. Universität Hamburg 1963/64, p.p. 111.
42. — Kennzeichnung kompakter Simplexe mit abgeschlossener Extremalpunktmenge. Arch. Math. **14**, 415—421 (1963).
43. — Darstellung von Bilinearformen auf Funktionenalgebren durch Integrale. Math. Z. **85**, 107—115 (1964).
44. — Aspects of linearity in the theory of function algebras. Sympos. on Function Algebras, Tulane 1965. Scott, Foresman and Co. 1966.
45. — Harmonische Räume und ihre Potentialtheorie. Lecture Notes in Math., Vol. 22. Berlin-Heidelberg-New York: Springer 1966.
46. — Choquetscher Rand und Integraldarstellungen. Jahresbericht, Deutscher Math. Ver. **69**, 89—104 (1967).
47. — Bear, H. S.: The part metric in convex sets (to appear).
48. — (cf. also Hinrichsen, D., Bauer, H.).
49. Bear, H. S.: The Šilov boundary for a linear space of continuous functions. Amer. Math. Monthly **68**, 484—485 (1961).
50. — A geometric characterization of Gleason parts. Proc. Amer. Math. Soc. **16**, 407—412 (1965).

51. Bauer, H.: Continuous supports for function spaces. Proc. Sympos. on Function Algebras, Tulane 1965, Scott, Foresman and Co. 1966.
52. — The integral representation of functions on parts. Illinois J. Math. **10**, 49—55 (1966).
53. — Lectures on Gleason parts. Lecture Notes in Mathematics **121**. Berlin-Heidelberg-New York: Springer 1970.
54. — Topologies of the Šilov boundary and equicontinuity in function spaces (to appear).
55. — The part metric in a cone (to appear: Pacific J. Math.).
56. — Walsh, B.: Integral kernel for one-part function spaces. Pacific J. Math. **23**, 209—215 (1967).
57. — Weiss, M.: An intinsic metric for parts. Proc. Amer. Math. Soc. **18**, 812—817 (1967).
58. — (cf. also Bauer, H., Bear, H. S.).
59. Bee Bednar, J.: On the Dirichlet problem for functions on the extreme boundary of a compact convex set (to appear in Math. Scand.).
60. Bessaga, C., Pelczynski A.: On extreme points in separable conjugate spaces. Israel J. Math. **4**, 262—264 (1966).
61. Birkhoff, G.: Lattice theory. Amer. Math. Soc. Coll. Publ. **25**, New York 1948.
62. Bishop, E.: A minimal boundary for function algebras. Pacific J. Math. **9**, 629—642 (1959).
63. — Leeuw, K. de: The representation of linear functionals by measures on sets of extreme points. Ann. Inst. Fourier (Grenoble) **9**, 305—331 (1959).
64. — Phelps, R. R.: The support functionals of a convex set. Proc. Sympos. in Pure Math. **7**, Convexity Amer. Math. Soc. 1963.
65. Björck, G.: The Set of extreme points of a compact convex set. Ark. Mat. **42**, 463—468 (1957).
66. Blackwell, D.: Comparison of experiments. Proc. of the Second Berkeley Symposium on Math. Stat. and Probability, Berkeley 1950, 93—102.
67. Blumenthal, R. M., Lindenstrauss, J., Phelps, R. R.: Extreme operators into $C(K)$. Pacific J. Math. **15**, 747—756 (1965).
68. Boboc, N.: Sur le noyaux sur un espace mesurable. Principe de domination. Rev. Roumaine Math. Pures Appl. **14**, 733—752 (1969).
69. — Cônes convexes de fonctions continues sur un espace compact. Rev. Roumaine Math. Pures Appl. **14**, 937—948 (1969).
70. — Bucur, G. H.: Cônes convexes ordonnés. Rev. Roumaine Math. Pures Appl. **14**, 283—309 (1969).
71. — Cornea, A.: Cônes des fonctions continues sur un espace compact. C. R. Acad. Sci. Paris **261**, 2564—2567 (1965).
72. — — Convex cones of lower semicontinuous functions on compact spaces. Rev. Roumaine Math. Pures Appl. **12**, 471—525 (1967).
73. — — Cônes convexes ordonnés. H-cônes et adjoints de H-cônes. C. R. Acad. Sci. Paris **270**, 596—599 (1970).
74. Bonenblust, H. F., Karlin, S.: Geometrical properties of the unit sphere of Banach algebras. Ann. of Math. **62**, 217—229 (1955).
75. Bonnesen, T., Fenchel, W.: Konvexe Körper. Ergebnisse der Math., Vol. 1. Berlin: Springer 1934.
76. Bucur, G. H. (cf. Boboc, N., Bucur, G. H.).
77. Bonsall, F. F.: Sublinear functionals and ideals in partly ordered vector spaces. Proc. London Math. Soc. **4**, 402—418 (1954).
78. — Endomorphisms of a partially ordered vector space without order unit. J. London Math. Soc. **30**, 144—153 (1955).

79. Bonsall, F. F.: Extreme maximal ideals in partially ordered vector spaces. Proc. Amer. Math. Soc. **7,** 831—837 (1956).
80. — Regular ideals of partially ordered vector spaces. Proc. London Math. Soc. **6,** 626—640 (1956).
81. — On the representation of points of a convex set. J. London Math. Soc. **24,** 265—272 (1963).
82. — Lindenstrauss, J., Phelps, R. R.: Extreme positive operators on algebras of functions. Math. Scand. **18,** 161—182 (1966).
83. Bourbaki, N.: Sur les espaces de Banach. C. R. Acad. Sci. Paris **206,** 1701—1704 (1938).
84. — Topologie gégérale Ch. I—II.
85. — Espaces vectoriels topologiques. Ch. I—II. Act. Sci. et Ind. **1189** (deuxième éd.), Paris 1966.
86. — Espaces vectoriels topologiques. Ch. III—V, Act. Sci. et Ind. **1229,** Paris 1955.
87. — Intégration, Ch. I, II, III, IV. Act. Sci. et Ind. **1175,** (deuxième éd.), Paris 1965.
88. — Intégration, Ch. V. Act. Sci. et Ind. **1244,** Paris 1956.
89. — Intégration. Ch. VI. Act. Sci. et Ind. **1281,** Paris 1959.
90. Bourgin, R. D.: Barycenters of measures on certain noncompact convex sets. Trans. Amer. Math. Soc. (to appear).
91. Brφndsted, A.: Conjugate convex functions in topological vector spaces. Mat.-Fys. Medd. Danske Vid. Selsk. **34,** (1964).
92. Buck, R. C.: A complete characterization for extreme functionals. Bull. Amer. Math. Soc. **65,** 130—133 (1959).
93. — Multiplication operators. Pacific J. Math. **11,** 95—104 (1961).
94. Bucy, R. S., Maltese, G.: A representation theorem for positive functionals on involutive algebras. Math. Ann. **162,** 364—367 (1966).
95. — — Extreme positive definite functions and Choquet's representation theorem. J. Math. Anal. Appl. **12,** 371—377 (1963).
96. Cartier, P., Fell, J. M. G., Meyer, P. A.: Comparaison des mesures portées par un unsemble convexe compact. Bull. Soc. Math. France **92,** 435—445 (1964).
97. Choquet, G.: Existence des représentations intégrales au moyen des points extrémeaux dans les cônes convexes. C. R. Acad. Sci. Paris **243,** 699—702 (1956).
98. — Existence des représentations intégrales dans les cônes convexes. C. R. Acad. Sci. Paris **243,** 736—737 (1956).
99. — Unicité des représentations intégrales au moyen des points extrémeaux dans les cônes convexes reticulés. C. R. Acad. Sci. Paris **243,** 555—557 (1956).
100. — Existence unicité des représentations intégrales au moyen des points extrémeaux dans les cônes convexes. Séminaire Bourbaki (Dec. 1956) 139, 15 pp.
101. — Le théorème de représentation intégrales dans les ensemble convexes compacts. Ann. Inst. Fourier (Grenoble) **10,** 333—344 (1960).
102. — Limites projectives d'ensemble convexes et éléments extrémeaux. C. R. Acad. Sci. Paris **250,** 2495—2497 (1960).
103. — Représentations intégrales dans les cônes convexes sans base compacte. C. R. Acad. Sci. Paris **253,** 1901—1903 (1961).
104. — Remarque à propos de la démonstration de l'unicité de P. A. Meyer. Séminaire Brelot-Choquet-Deny (Théorie de Potentiel), 6e année (1961/62), No. 8.
105. — Ensembles et cônes convexes faiblement complets. (Two papers.) C. R. Acad. Sci. Paris **254,** 1908—1910, 2123—2125 (1962).
106. — Axiomatique des mesures maximales. Application aux cônes convexes faiblement complets. R. C. Acad. Sci. Paris **255,** 37—39 (1962).
107. — Étude des mesures coniques, cônes convexes saillant faiblement complets sans genératrices extrémales. C. R. Acad. Sci. Paris **255,** 445—447 (1962).
108. — Cônes convexes faiblement complets. Internat. Congr. of Math., Stockholm 1962.

109. Choquet, G.: Lectures on analysis. Edited by J. Marsden, T. Lance and S. Gelbart. New York: Benjamin, Inc. 1969.
110. — Corson, H., Klee, V.: Exposed points of convex sets. Pacific J. Math. **17,** 33—43 (1966).
111. — Meyer, P. A.: Existence et unicité des représentations intégrales dans les convexes compacts quelconque. Ann. Inst. Fourier (Grenoble) **13,** 139—154 (1963).
112. Combes, J. F.: Sur les faces d'une C^*-algèbre. Bull. Sci. Math. 2e série, **93**, 37—62 (1969).
113. — Perdrizet, F.: Certains idéaux dans les espaces vectoriels ordonnés. C. R. Acad. Sci. Paris **268,** 1552—1555 (1969).
114. — — Certains idéaux dans les espaces vectoriels ordonnés. J. Math. Pures Appl. **49,** 29—59 (1970).
115. Converse, G., Namioka, I., Phelps, R. R.: Extreme invariant positive operators. Trans. Amer. Math. Soc. **137,** 375—385 (1969).
116. — Extreme positive operators on $C(X)$ which commute with certain operators. Trans. Amer. Math. Soc. (to appear).
117. Cornea, A.: (cf. Boboc, N., Cornea).
118. Corson, H.: Metrizability of compact convex sets from theorems on capacities and maximal measures. Amer. Math. Soc. (to appear).
119. — Lindenstrauss, J.: On weakly compact subsets of Banach spaces. Proc. Amer. Math. Soc. **17,** 407—412 (1966).
120. — (cf. also Choquet, G., Corson, Klee, V.).
121. Dantzer, L., Grünbaum, B., Klee, V.: Helly's Theorem and it relatives. Proc. of Symposia in Pure Math., vol. **7,** Convexity, Amer. Math. Soc. 101—180 (1962).
122. Day, M. M.: Normed linear spaces. Ergebnisse Math. Berlin-Göttingen-Heidelberg: Springer 1958.
123. Dauns, F., Hofmann, K. H.: Representations of rings by continuous sections. Mem. Amer. Math. Soc. **83** (1968).
124. Davies, E. B.: A generalized theory of convexity. Proc. London Math. Soc. **17,** 644—652 (1967).
125. — On the Banach duals of certain spaces with the Riesz decomposition property. Quart. J. Math. Oxford **18,** 399—414 (1967).
126. — The structure and ideal theory of the predual of a Banach lattice. Trans. Amer. Math. Soc. **131,** 544—555 (1968).
127. — On the Borel structure of C^*-algebras. Comm. Math. Phys. **8,** 147—163 (1968).
128. — Decomposition of traces on separable C^*-algebras. Quart. J. Math. Oxford **20,** 97—111 (1969).
129. — The structure of $\sum^*$-algebras. Quart. J. Math. Oxford **20,** 351—366 (1969).
130. — Vincent-Smith, G. F.: Tensor products, infinite products, and projective limits of Choquet simplexes. Math. Scand. **22,** 145—164 (1968).
131. Dieudonné J.: La dualité dans les espaces vectoriels topologiques. Ann. Sci. École Norm. Sup. Paris **59,** 129 (1942).
132. — Natural homomorphisms in Banach spaces. Proc. Amer. Math. Soc. **1,** 54—59 (1959).
133. — Sur la séparation des ensembles convexes. Math. Ann. **163,** 1—3 (1966).
134. — Schwartz, L.: Sur les espaces $\mathscr{L}$ et $\mathscr{LF}$. Ann. Inst. Fourier (Grenoble) 61—101 (1949).
135. Dixmier, J.: Sur un théorème de Banach. Duke Math. J. **15,** 1057—1071 (1948).
136. — Les algèbres d'opérateurs dans l'espace hilbertien. Paris: Gauthier-Villars 1957.
137. — Sur les C^*-algèbres. Bull. Soc. Math. France **88,** 95—112 (1960).
138. — Les C^*-algèbres et leurs représentations. Paris: Gauthier-Villars 1964.
139. Douglas, R. G.: On extremal measures and subspace density. Michigan Math. J. **11,** 644—652 (1964).

140. Dunford, Schwartz: Linear operators Part I. New York 1958.
141. Edwards, D. A.: On the representation of certain functionals by measures on the Choquet boundary. Ann. Inst. Fourier (Grenoble) **13,** 111—121 (1963).
142. — The homeomorphic affine embedding of a locally compact cone into a Banach dual space endowed with the vague topology. Proc. London Math. Soc. **14,** 399—414 (1964).
143. — Séparation de fonctions réelles définies sur un simplexe de Choquet. C. R. Acad. Sci. Paris **261,** 2798—2800 (1965).
144. — Minimum stable wedges of semi-continuous functions. Math. Scand. **19,** 15—26 (1966).
145. — On separation and approximation of real functions defined on a Choquet simplex. Proc. Second Prague Topological Symposium. 122—128 (1966).
146. — A class Choquet boundaries that are Baire spaces. Quart. J. Math. Oxford **17,** 282—284 (1966).
147. — The affine continuous functions on a Choquet simplex. Proc. Bruges Summer School on Topological Algebra Theory (1966), Brussels 1967.
148. — Choquet boundary theory for certain spaces of lower semi-continuous functions. Proc. Tulane Conf. on Function Algebras, 1965, Scott, Foresman and Co. (1966).
149. — An extension of Choquet boundary theory to certain partially ordered compact convex sets. Studia Math. **36,** 177—193 (1970).
150. — On separation and approximation of real functions defined on a Choquet simplex. Symposium on general topology ect. Praha, 122—128 (1966).
151. — On uniform approximation of affine functions on a compact convex set. Quart. J. Math. Oxford **78,** 139—142 (1969).
152. — Vincent-Smith, G.: A Weierstrass-Stone theorem for Choquet simplexes. Ann. Inst. Fourier (Grenoble) **18,** 261—282 (1968).
153. Effros, E. G.: A decomposition theorem for representations of C^*-algebras. Trans. Amer. Math. Soc. **107,** 83—106 (1963).
154. — Order ideals in a C^*-algebra and its dual. Duke Math. J. **30,** 391—412 (1963).
155. — Structure in simplexes. Acta Math. **117,** 103—121 (1967).
156. — Structure in simplexes II. J. of Functional Analysis **1,** 361—391 (1967).
157. — Gleit, Alan: Structure in simplexes III, Composition series. Trans. Amer. Math. Soc. **142,** 355—379 (1969).
158. — Structure in simplexes IV (to appear).
159. — (cf. also Aarnes, J. F., Effros, E. G., Nielsen, O. A.).
160. — Kazdan, J.: Applications of Choquet simplexes to elliptic and parabolic boundary value problems. J. Differential Equations **8,** 95—134 (1970).
161. — — On the Dirichlet problem of the heat equation (to appear, J. Math. Mech.).
162. Eggleston, H. G.: Convexity. Cambridge Tracts in Math. and Math. Phys. **47,** Cambridge University Press 1958.
163. Ellis, A. J.: Extreme positive operators. Quart. J. Math. Oxford **15,** 342—344 (1964).
164. — The duality of partially ordered normed linear spaces. J. London Math. Soc. **39,** 730—744 (1964).
165. — Perfect order ideals. J. London Math. Soc. **40,** 288—294 (1965).
166. — Linear operators in partially ordered normed vector spaces. J. London Math. Soc. **41,** 323—332 (1966).
167. — An intersection property for state spaces. J. London Math. Soc. **43,** 173—176 (1968).
168. — On partial orderings of normed spaces. Math. Scand. **23,** 123—132 (1968).
169. — On faces of compact convex sets and their annihilators. Math. Ann. **184,** 19—24 (1969).
170. — Lecture notes on affine functions and faces of convex sets. Calif. Inst. of Tech. pp. 46 (mimeographed).

171. Ellis, A.J.: Facial decomposition and a separation property of linearly compact simplexes (to appear).
172. Fakhoury, H.: Solution d'un problème posé par Effros. C. R. Acad. Sci. Paris **269,** 77—79 (1969).
173. — Caractérisation des simplexes compacts. C. R. Acad. Sci. Paris **269,** 21—24 (1969).
174. — Stabilité des simplexes de Lion. C. R. Acad. Sci. Paris **270,** 110—112 (1970).
175. — Une caractérisation des simplexes compats et des cônes réticulés. Applications. Séminaire Choquet (Initiation à l'Analyse) 9e année, 1969/70, no. 2. (mimeographed pp. 12).
176. — Goullet de Rugy, A., Rogalski, M.: Centre d'un M-espace. C. R. Acad. Sci. Paris **270,** 1744—1748 (1970).
177. Fan, Ky: On the Krein-Milman theorem. Proc. of Symposia in Pure Math. Convexity. Amer. Math. Soc. **7,** 211—219 (1963).
178. Fell, J. M. G.: The dual spaces of C^*-algebras. Trans. Amer. Math. Soc. **94,** 365—403 (1960).
179. — (cf. Cartier, P., Fell, J. M. G., Meyer, P. A.).
180. Fenchel, W.: On conjugate convex functions. Canad. J. Math. **1,** 73—77 (1949).
181. — Convex cones, sets and functions. Lecture notes. Princeton, N.J.: Princeton University 1953 (mimeographed).
182. — (cf. also Bonnesen, Fenchel).
183. Gelfand, I. M., Raikov, D. A., Silov, G. E.: Commutative normed rings. Uspehi Mat. Nauk. **2** (N.S.), 48—146 (1946). (In Russian. Amer. Math. Soc. Translation Series 2, Vol. 5, 1957, 115—220).
184. Gleason, A. M.: Function algebras. Seminar on Analytic Functions. Vol. 2. Inst. Adv. Study Princeton N.J. 1956.
185. — The abstract theorem of Cauchy-Weil. Pacific J. Math. **12,** 511—525 (1962).
186. Gleit, Alan S.: Topics in simplex space theory. Dept. of Math. Stanford Univ. 1969 (mimeographed, pp. 92).
187. Glimm, J.: Type I C^*-algebras. Ann. of Math. **13,** 572—612 (1961).
188. Goullet de Rugy, A.: Caractère réticulé de certains cônes de fonctions linéaire sur un cône convexe décomposable. Séminaire Brelot-Choquet-Deny (Théorie de Potentiel) **12,** (1967/68) no. 5, Paris 1968.
189. — Faces complémentables dans un simplex. Séminaire Brelot-Choquet-Deny (Théorie du Potentiel) **12,** (1967/68) no. 6, Paris 1968.
190. — Étude des faces complémentables dans un simplexe compact. C. R. Acad. Sci. Paris **267,** 736—739 (1968).
191. — Quelques proprietés des fonctions numériques linéaires et semi-continues sur un cône convexe saillant faiblement complet. C. R. Acad. Sci. Paris **265,** 841—844 (1967).
192. — Géometrie des simplexes. Centre de Documentation Universitaire, Paris 1968, pp. 84.
193. — Étude de quelques espaces de fonctions affines sur un simplexe. C. R. Acad. Sci. Paris **267,** 640—643 (1968).
194. — Faces parallèlisables et topologies faciales sur l'espace des états d'un algèbre stellaire. C. R. Acad. Sci. Paris **270,** 376—379 (1970).
195. — Invitation à la théorie des algèbres stellaire. Séminaire Choquet (Initation à l'analyse) 9e année 1969/70; no. 4,5.
196. — (cf. also Aljani, M., Goullet de Rugy, A.).
197. — (cf. also Fakhoury, H., Goullet de Rugy, A., Rogalski, M.).
198. Grosberg, J., Krein, M.: Sur la décomposition des fonctionelles en composantes positives. C. R. (Doklady) de l'Acad. Sci. de l'URSS **25,** 723—726 (1939).
199. Grossman, M. W.: A Choquet boundary for the product of two compact spaces. Proc. Amer. Math. Soc. **16,** 967—977 (1965).

200. Grossman, M. W.: Relations of a paper of Ky Fan to a theorem of Krein-Milman type. Math. Z. **90,** 212—214 (1965).
201. — The Choquet boundary and extreme sets. Math. Z. **95,** 259—271 (1967).
202. — Relative Choquet and Silov boundaries. J. Reine Angewandte Math. **225,** 1—29 (1967).
203. — An extremal property for certain spaces of lower semi-continuous functions. Math. Z. **106,** 139—148 (1968).
204. Grothendieck, A.: Sur la completion du dual d'un espaces vectoriel localement convexe. C. R. Acad. Sci. Paris **230,** 605—606 (1950).
205. — Une caractérisation vectorielle métrique des espaces L^1. Canad. J. Math. **7,** 552—561 (1955).
206. Grünbaum, B.: Convex polytopes. Interscience Publishers New York, London, 1967.
207. — (cf. also Dantzer, L., Grünbaum, B., Klee, V.).
208. Hardy, G. H., Littlewood, J. E., Polyá, G.: Inequalities. 2nd. ed. Cambridge University Press, 1952.
209. Hervé, M.: Sur les représentations intégrales à l'aide des points extrémeaux dans un ensemble compact convexe métrisable. C. R. Acad. Sci. Paris **253,** 366—368 (1961).
210. Hewitt, E.: Integral representation of certain linear functionals. Ark. Mat. **2,** 269—282 (1952).
211. Hinrichsen, D.: Adapted integral representations by measures on Choquet boundaries. Bull. Amer. Math. Soc. **72,** 888—891 (1966).
212. — Bauer, H.: Einige Eigenschaften lokalkompakter konvexer Mengen und ihrer projektiven Limiten. Colloquium on Convexity, Copenhagen 1965. Edited by Math. Inst. Univ. of Copenhagen, 143—153 (1967).
213. Hirsberg, B.: (cf. Alfsen, E. M., Hirsberg, B.).
214. Hoffman, K.: Minimal boundaries for analytic polyhedra. Rend. Circ. Mat. Palermo **9,** 147—160 (1960).
215. Hofmann, K. H.: (cf. Dauns, F., Hofmann, K. H.).
216. Hulanicki, A., Phelps, R. R.: Some applications of tensor products of partially ordered linear spaces. J. Functional Analysis **2,** 177—201 (1968).
217. Husain, T.: The open mapping and closed graph theorems in topological vector spaces. Oxford 1965.
218. Hustad, O.: Convex cones with properties related to weak local compactness. Math. Scand. **11,** 79—90 (1962).
219. — Extension of positive linear functionals. Math. Scand. **11,** 63—78 (1962).
220. — Duality theorems of Minkowski type. Colloquium on Convexity. Copenhagen 1965. Edited by Math. Inst. Univ. of Copenhagen, 154—164 (1967).
221. Jellet, F.: Homomorphisms and inverse limits of Choquet simplexes. Math. Z. **103,** 219—226 (1968).
222. Johansen, S.: An application of extreme point methods to the representation of infinitely divisible distributions. Z. Wahrscheinlichkeitstheorie und Verw. Gebiete **5,** 304—316 (1966).
223. — Applications of extreme point methods in probability. Inst. Mat. Stat. Copenhagen Univ. 1967 (p. 79, mim. in Danish).
224. Kadison, R. V.: A representation theory for commutative topological algebra. Mem. Amer. Math. Soc. **7** (1951).
225. — Order properties of bounded self-adjoint operators. Proc. Amer. Math. Soc. **2,** 505—510 (1951).
226. — Unitary invariants for representations of operator algebras. Ann. of Math. **66,** 304—379 (1957).
227. — States and representations. Amer. Math. Soc. **103,** 304—319 (1962).
228. — Transformation of states in operator theory and dynamics. Topology 3, Supplement 2, 177—198 (1965).

229. Kadison, R. V.: Operator algebras with a faithful weakly closed representation. Ann. of Math. **64**, 175—181 (1956).
230. — A generalized Schwarz inequality and algebraic invariants for operator algebras. Ann. of Math. **56**, 494—503 (1952).
231. — Isometries of operator algebras. Ann. of Math. **54**, 325—338 (1951).
232. — Singer, I. M.: Extensions of pure states. Amer. J. Math. **81**, 383—400 (1959).
233. Kakutani, S.: Weak topology and regularity of Banach spaces. Proc. Imp. Acad. Tokyo **15**, 169—173 (1939).
234. — Concrete representations of abstract L-spaces and the mean ergodic theorem. Ann. of Math. **42**, 523—537 (1941).
235 — Concrete representations of abstract (M)-spaces. Ann. of Math. **42**, 994—1024 (1941).
236. Kaplansky, I.: The structure of certain operator algebras. Trans. Amer. Math. Soc. **70**, 219—255 (1952).
237. Karlin, S.: (cf. Bohnenblust, H. F., Karlin, S.).
238. Kazdan, J.: (cf. Effros, E. G., Kazdan, J.).
239. Kelley, J. L.: General topology. Princeton N.J.: Van Nostrand Co. 1955.
240. — Namioka, I.: Linear topological spaces. Princeton N.J.: Van Nostrand Co. 1963.
241. — Vaught, R.: The positive cone in Banach algebras. Trans. Amer. Math. Soc. **74**, 44—55 (1953).
242. Kendall, D. G.: Simplexes and vector lattices. J. London Math. Soc. **37**, 365—371 (1962).
243. Klee, V.: Convex sets in linear spaces. Duke Math. J. **18**, 443—466 (1951).
244. — Convex sets in linear spaces II. Duke Math. J. **18**, 875—883 (1951).
245. — Separation properties of convex cones. Proc. Amer. Math. Soc. **6**, 313—318 (1955).
246. — Extremal structure of convex sets. Arch. Math **8**, 234—240 (1957).
247. — Extremal structure of convex sets II. Math. Z. **69**, 90—104 (1958).
248. — Relative extreme points. Proc. Internat. Symposium on Linear Spaces. Jerusalem 1960, 282—289.
249. — Infinite dimensional intersection theorem. Proc. Symposia in Pure Math. **7**, Convexity. Amer. Math. Soc. 1963.
250. — (cf. also Dantzer, L., Grünbaum, B., Klee, V.).
251. — (cf. also Choquet, G., Carson, H., Klee, V.).
252. Krein, M., Krein, S.: On an inner characteristic of the set of all continuous functions defined on a bicompact Hausdorff space. C. R. (Doklady) Acad. Sci. URSS **27**, 427—430 (1940).
253. — Milman, D.: On the extreme points of regularly convex sets. Studia Math. **9**, 133—138 (1940).
254. — Smullyan, V.: On regularly convex sets in the space conjugate to a Banach space. Ann. of Math. **41**, 556—583 (1940).
255. Krause, U.: Der Satz von Choquet als ein abstrakter Spektralsatz und vice versa. Math. Ann. **184**, 275—296 (1970).
256. Köthe, G.: Topologische lineare Räume I. Grundlagen der Math. Wiss. **107**, Berlin-Göttingen-Heidelberg: Springer 1960.
257. Lanford, O., Ruelle, D.: Integral representations of invariant states on B^*-algebras. J. Math. and Phys. **8**, 1460—1463 (1967).
258. Lazar, A.: Spaces of affine functions on simplexes. Hebrew University 1965 (mimeographed).
259. — Affine functions on simplexes and extreme operators. Israel J. Math. **5**, 31—43 (1967).
260. — Affine products of simplexes. Math. Scand. **22**, 165—175 (1968).

261. Lazar, A.: Spaces of affine continuous functions on simplexes. Amer. Math. Soc. Trans. **134,** 503—525 (1968).

262. — Polyhedral Banach spaces and extensions of compact operators. Israel J. Math. **7,** 357—364 (1969).

263. — Sections and subsets of simplexes (to appear).

264. — The unit ball of conjugate L^1-spaces. Duke Math. J. (to appear).

265. — Lindenstrauss, J.: On Banach spaces whose duals are L_1-spaces. Israel J. Math. **4,** 205—207 (1966).

266. — — Banach spaces whose duals are L_1-spaces and their representing matrices (to appear).

267. — Wulbert, D. E.: Continuous selections for metric projections. J. Functional Analysis **3,** 193—216 (1969).

268. Lacey, H. E., Morris, P. D.: On spaces of type $A(K)$ and their duals. Proc. Amer. Math. Soc. **23,** 151—157 (1969).

269. Leeuw, K. de: (cf. Bishop, E., Leeuw, K. de).

270. — (cf. Rudin, W., Leeuw, K. de).

271. Léger, C.: Une démonstration du théorème de A. J. Lazar sur les simplexes compacts. C. R. Acad. Sci. Paris **265,** 830—831 (1967).

272. — Soury, P.: Le convexe topologique des probabilités sur un espace topologique. C. R. Acad. Sci. Paris **270,** 516—518 (1970).

273. Lindenstrauss, J.: Extension of compact operators. Mem. Amer. Math. Soc. **48,** Providence R. I. (1964).

274. — On operators which attain their norm. Israel J. Math. **3,** 139—148 (1963).

275. — On extreme points in l_1. Israel J. Math. **4,** 59—61 (1966).

276. — Weakly compact sets,—their topological properties, and the Banach spaces they generate. Proc. of the Symposium of Infinite Dimensional Topology, Baton Rouge 1966 (to appear).

277. — Phelps, R. R.: Extreme point properties of convex bodies in reflexive Banach spaces. Israel J. Math. **6,** 39—48 (1968).

278. — Wulbert, D. E.: On the classification of the Banach spaces whose duals are L_1 spaces. J. Functional Analysis **4,** 332—349 (1969).

279. — (cf. also Amir, D. Lindenstrauss, J.).

280. — (cf. also Lazar, A. Lindenstrauss, J.).

281. — (cf. also Corson, H. H., Lindenstrauss, J.).

282. Loomis, L. H.: Unique direct integral decompositions on convex sets. Amer. J. Math. **94,** 509—526 (1962).

283. Lumer, G.: Points extrémeaux associés; frontières de Shilov et Choquet; principe du minimum. C. R. Acad. Sci. Paris **256,** 858–861 (1963).

284. — Points extreméaux associés; applications aux cônes de fonctions semicontinues. C. R. Acad. Sci. Paris **256,** 1066—1068 (1963).

285. Malthese, G.: (cf. Bucy, R. S., Malthese, G.).

286. Maserick, P. H.: Dual convex polytopes in Banach spaces. J. Math. Anal. Appl. **19,** 263—273 (1967).

287. Mazur, S.: Über konvexe Mengen in linearen normierten Räumen. Studia Math. **4,** 70—84 (1933).

288. Meyer, P. A.: Sur les démonstrations nouvelles du théorème de Choquet. Seminaire de Théorie du Potential (Brelot, Choquet, Deny) Paris 1961/62. Fasc. **2,** Exposé 7, page 07.

289. — Probability and Potentials. Waltham, Mass.: Blaisdell 1966.

290. — (cf. also Cartier, P., Fell, J. M. G., Meyer, P. A.).

291. Michael, E.: Continuous selections I. Ann. of Math. **63,** (1956).

292. — Pełczynski, A.: Separable Banach spaces which admit l_n^∞ approximations. Israel J. Math. **4,** 189—198 (1966).

293. Miller, R.R.: Gleason parts and Choquet boundaries in convolution measures algebras. Pacific J. Math. **31,** 755—771 (1969).

294. Milman, D.P.: Characteristics of extremal points of regularly convex sets. Dokl. Akad. Nauk SSSR **57,** 119—122 (1947).

295. — Rutman, M.A.: On a more precise theorem about the completeness of the system of extremal points of a regularly convex set. Dokl. Akad. Nauk. SSSR **60,** 25—27 (1948).

296. — The facial structure of a convex bicompact space and integral decomposition of means. Dokl. Akad. Nauk. SSSR **88,** 357—360 (1952).

297. Mokobodzki, G.: Balayage défini par un cône convexe de fonctions numériques sur un espace compact. C.R. Acad. Sci. Paris **254,** 803—805 (1962).

298. — Principe de balayage, principe de domination. Séminaire Choquet **1,** no. 1 (1962).

299. — Quelques proprietés des fonctions numèriques convexes (s.c.i. on s.c.s.) sur un ensemble convexe compact. Séminaire Brelot-Choquet-Deny (Théorie du Potentiel) 6e année no. 9 (1962).

300. — Barycentres généralisées. Séminaire Brelot-Choquet-Deny **6,** no. 13 (1962).

301. — Sibony, D.: Cônes de fonctions continues. C.R. Acad. Sci. Paris **264,** 15—18 and 238—241 (1967); **265,** 21—24 (1967); **266,** 215—218 (1968).

302. Morris, P.D., Phelps, R.R.: Theorems of Krein-Milman type for certain convex sets of operators. Trans. Amer. Math. Soc. (to appear).

303. — — Theorems of Krein-Milman type for certain convex sets of functions and operators. Ann. Inst. Fourier (to appear).

304. — (cf. also Lacey, H.E., Morris, P.D.).

305. Nachbin, L.: A characterization of the normed vector ordered spaces of continuous functions over a compact space. Amer. J. Math. **71,** 701—705 (1949).

306. — Sur l'abondance des points extrémaux d'un ensemble convexe borné et fermé. Ann. Acad. Brasil Ci. **34,** 445—448 (1962).

307. — Topology and order. New York: Van Nostrand 1965.

308. Nagel, R.: Idealtheorie in geordneten lokalkonvexen Vektorräumen. Eberhard-Karls-Universität zu Tübingen. Thesis pp. 53, (1969).

309. — Ideals in ordered locally convex spaces. Math. Scand. (to appear).

310. Namioka, I.: Partially ordered linear topological spaces. Mem. Amer. Math. Soc. **24,** (1957).

311. — Neighbourhoods of extreme points. Israel J. Math. **5,** 145—152 (1967).

312. — (cf. also Converse, G., Namioka, I., Phelps, R.R.).

313. Nielsen, O.A.: (cf. also Aarnes, J.F., Effros, E.G., Nielsen, O.A.).

314. Ng, Kung-Fu: A note on simplex spaces. Proc. Cambridge Philos. Soc. **66,** 559—562 (1969).

315. — The duality of partially ordered Banach spaces. Proc. London Math. Soc. **19,** 269—288 (1969).

316. Nordseth, T.: (cf. also Alfsen, E.M., Nordseth, T.).

317. Ore, Ø.: On the foundations of abstract algebra. Ann. Math. **36,** 406—437 (1935).

318. Pedersen, G. Kjærgard: A decomposition theorem for C*-algebras. Math. Scand. **22,** 266—268 (1968).

319. Perdrizet, F.: Sur certains espaces de Banach ordonnés. Bull. Sci. Math. 2e série **92,** 129—141 (1968).

320. — Espaces de Banach ordonnés et idéaux. C.R. Acad. Sci. Paris **269,** 393—396 (1969).

321. — Espaces de Banach ordonnés et idéaux. J. Math. Pures Appl. **49,** 61—98 (1970).

322. — Idéaux dans les espaces de type (*F*) (to appear).

323. — (cf. also Combes, F., Perdrizet, F.).

324. Pełczynski, A.: (cf. also Michael, E., Pełczynski, A.).

325. — (cf. also Bessaga, C., Pełczynski, A.).

326. Peressini, A. L.: Ordered topological vector spaces. New York: Harper & Row 1967.
327. Phelps, R. R.: Representation theorems for bounded convex sets. Proc. Amer. Math. Soc. **11**, 976—983 (1960).
328. — Support cones and their generalizations. Sympos. in Pure Math., Vol. 7, Convexity. Amer. Math. Soc. 1963.
329. — Weak* support points of convex sets in E^*. Israel J. Math. **2**, 177—181 (1964).
330. — Extreme points in function algebras. Duke Math. J. **32**, 267—277 (1965).
331. — Lectures on Choquet's theorem. Math. Studies. Princeton: Van Nostrand 1966.
332. — Infinite dimensional compact convex polytopes. Math. Scand. **24**, 5—26 (1969).
333. — (cf. also Bishop, E., Phelps, R. R.).
334. — (cf. also Converse, G., Namioka, I., Phelps, R. R.).
335. — (cf. also Hulanicki, A., Phelps, R. R.).
336. — (cf. also Lindenstrauss, J., Phelps, R. R.).
337. — (cf. also Morris, P. D., Phelps, R. R.).
338. Poulsen, Ebbe Thue: A simplex with dense extreme points. Ann. Inst. Fourier (Grenoble) **11**, 83—87 (1961).
339. Price, G. B.: On the extreme points of convex sets. Duke Math. J. **3**, 54—67 (1937).
340. Prosser, R.: On the ideal structure of operator algebras. Mem. Amer. Math. Soc. **45**, 1—28 (1963).
341. Raikov, D. A.: (cf. also Gelfand, I. M., Raikov, D. A., Šilov, G. E.).
342. Riesz, F.: Sur quelques notions fondamentales dans la théorie générale des operations linéaires. Ann. of Math. **41**, 174—206 (1940).
343. Rogalski, M.: Étude du quotient d'un simplexe par une face fermée; application à un théorème de Alfsen. Séminaire Brelot-Choquet-Deny (Théorie du Potentiel) **12** (1967/68) no. 2, Paris 1967.
344. — Espaces de Banach ordonnés, simplexes, frontières de Šilov et problème de Dirichlet. Séminaire Choquet (Initiation à l'Analyse), 5e année (1965—66), no. 12.
345. — Représentations fonctionelles d'espaces vectoriels réticulés. Séminaire Choquet. (Initiation à l'Analyse), 5e année (1965—66), no. 2.
346. — Opérateurs de Lion, projecteurs boréliens, et simplexes analytiques. Service des Publications, Fac. des Sci. Orsay (France), 1967—68.
347. — Topologie faciale dans les convexes compacts. C. R. Acad. Sci. Paris **270**, 523—526 (1970).
348. — Topologies faciales dans les convexes compacts. C. R. Acad. Sci. Paris **270**, 766—768 (1970).
349. — Calcul fonctionel et décomposition spectrale dans le centre d'un espace $A(X)$. C. R. Acad. Sci. Paris **270**, 820—823 (1970).
350. — Calcul fonctionel et décomposition spectrale pour une fonction facialement semi-continues sur un convexe compact. C. R. Acad. Sci. Paris **270**, 868—871 (1970).
351. — Caractérisation des simplexes par des proprietés portant sur les faces fermées et sur les ensembles compacts de points extrémeaux (to appear in Math. Scand.).
352. — (cf. also Fakhoury, H., Goullet de Rugy, A., Rogalski, M.).
353. Rudin, W., Leeuw, K. de: The extreme points of the unit ball in H_1. Amer. Math. Soc. Notices **5**, 177 (1958).
354. Ruelle, D.: Integral representations of states on a C^*-algebra (to appear).
355. — (cf. also Lanford, O., Ruelle, D.).
356. Russo, B.: (cf. also Akemann, C. A., Russo, B.).
357. Sakai, S.: On the central decomposition for positive functionals on C^*-algebras. Trans. Amer. Math. Soc. **118**, 406—419 (1965).
358. Schaeffer, H. H.: Invariant ideals of positive operators in $C(X)$. I. Illinois J. Math. **11**, 703—715 (1967).
359. — Invariant ideals of positive operators in $C(X)$. II. Illinois J. Math. **12**, 525—538 (1968).

360. Schwartz, L.: (cf. also Dieudonné, D. S., Schwartz, L.).
361. Semadeni, Z.: Free compact convex sets. Bull. Acad. Sci. Pol. **13,** 141—146 (1965).
362. — Vector lattices. Lecture Notes Series Århus University 1965 (pp. 49, mimeographed).
363. — Categorial methods in convexity. Proc. Coll. on Convexity, Copenhagen 281—307 1965. Edited by Math. Inst. Copenhagen University 1967.
364. Sibony, D.: Cônes des fonctions et potentiels. Cours de 3ème Cycle, Fac. des Sci. de Paris 1967—68 (mimeographed pp. 150).
365. — Frontières associées à des cônes convexes de fonctions s.c.i. Contrib. Extens. Theory of Topol. Struct. Proc. Sympos. Berlin 1967, 199—206 (1969).
366. — (cf. also Mokobodzki, G., Sibony, D.).
367. Šilov, G. E.: (cf. also Gelfand, I. M., Raikov D. A., Šilov, G. E.).
368. Siciak, J.: On function families with boundary. Pacific J. Math. **12,** 375—384 (1962).
369. Sherman, S.: On the theorem of Hardy, Littlewood, Polyá and Blackwell. Proc. Nat. Acad. Sci. USA **37,** 826—831 (1951).
370. Singer, I. M.: (cf. Arens, R., Singer, I. M.).
371. Skau, C. F.: Existence of simplicial boundary measures on compact convex sets. Aarhus Univ. Various Publ. Series, no 5, 1969 (p. 66, mimeographed).
372. Soury, P.: (cf. also Léger, C., Soury, P.).
373. Stefánson, J.: On a problem of J. Dixmier concerning ideals in a von Neumann algebra. Math. Scand. **24,** 111—112 (1969).
374. Šmulian, V.: (cf. also Krein, M., Šmulian, V.).
375. Stone, M. H.: A general theory of spectra I. Proc. Nat. Acad. Sci. USA **26,** 280—283 (1940).
376. — A general theory of spectra II. Proc. Nat. Acad. Sci. USA **27,** 83—87 (1941).
377. Strassen, V.: The existence of probability measures with given marginals. Ann. Math. Statist. **36,** 423—439 (1965).
378. Straszewicz, S.: Über exponierte Punkte abgeschlossener Punktmengen. Fund. Math. **24,** 139—143 (1935).
379. Störmer, E.: Positive linear maps of operator algebras. Acta Math. **110,** 233—278 (1963).
380. — On extremal maps of operator algebras. Århus Univ. January 1966.
381. — Two sided ideals in C^*-algebras. Amer. Math. Soc. Bulletin **73,** 254—257 (1967).
382. — On partially ordered vector spaces and their duals with applications to simplexes and C^*-algebras. Proc. London Math. Soc. (3) **18,** 245—265 (1968).
383. — A characterization of pure states of C^*-algebras. Proc. Amer. Math. Soc. **19,** 1100—1102 (1968).
384. Taylor, P. D.: The structure space of a Choquet simplex (to appear).
385. Thoma, E.: Über unitäre Darstellungen abzählbarer, diskreter Gruppen. Math. Ann. **153,** 111—138 (1964).
386. Topping, D. M.: Extreme points in (M)-spaces. Amer. Math. Soc. Notices **9,** (1962), **148,** Abstract 62 T—79.
387. — Extreme points in order intervals of self-adjoint operators. Amer. Math. Soc. Notices 9 (1962), 211. Abstract 62 T—113.
388. Tukey, J. W.: Some notes on the separation of convex sets. Portugal. Math. **3,** 95—102 (1942).
389. Valentine, A.: Convex sets. New York: McGraw Hill 1964.
390. Varadarajan, V. S.: Groups of automorphisms of Borel spaces. Trans. Amer. Math. Soc. **109,** 191—220 (1963).
391. Vaught, R.: (cf. also Kelley, J. L., Vaught, R.).
392. Vincent Smith, G. F.: A Choquet boundary theory for measures taking values in a Stone algebra. J. London Math. Soc. **44,** (1969).

393. — Positive maps between partially ordered bimodules. Proc. London Math. Soc. **19,** 661—674 (1969).
394. Vincent Smith, G. F.: An extension of Carathéodory's theorem to infinite dimensions using Choquet boundary theory (to appear).
395. — (cf. also Davies, E. B., Vincent-Smith, G. F.).
396. — (cf. also Edwards, D. A., Vincent-Smith, G. F.).
397. Watanabe, Takesi: Boundaries of Markov processes and related topics. Århus University, Lecture Notes 1969/70.
398. Walsh, B.: (cf. also Bear, H. S., Walsh, B.).
399. Weiss, M.: (cf. also Bear, H. S., Weiss, M.).
400. Wils, W.: Désintégration centrale des formes positives sur les C^*-algèbres. C. R. Acad. Sci. Paris **267,** 810—812 (1968).
401. — Désintégration centrale dans une partie convexe compacte d'un espace localement convexe. C. R. Acad. Sci. Paris **269,** 702—704 (1969).
402. — The ideal center of partially ordered vector spaces (to appear in Acta Math.).
403. Wulbert, D. E.: (cf. also Lazar, A., Wulbert, D. E.).
404. — (cf. also Lindenstrauss, J., Wulbert, D. E.).
405. Yosida, K.: On vector lattice with a unit. Proc. Imp. Acad. Tokyo **17,** 121—124 (1941).
406. Alfsen, E. M.: Un théorème de Weierstrass-Stone pour les sous espaces vectoriels de $C_R(X)$. C. R. Acad. Sci. Paris **271,** 725—726 (1970).
407. Björk, J. E.: Interpolation in closed linear subspaces of $C(X)$ (to appear).
408. Briem, E.: Restrictions of subspaces of $C(X)$. Aarhus Univ. Preprint Series no. 26 (1969/70).
409. Choquet, G.: Formes linéaires positives sur les espaces de fonctions. Espaces sous-stonien et pseudo-mesure. C. R. Acad. Sci. Paris **271**, 164—167 (1970).
410. — Formes linéaires positives sur les espaces de fonctions. Pseudo-mesures; éléments extrémaux; théorèmes de densité. C. R. Acad. Sci. Paris **271**, 828—831 (1970).
411. Cunningham, F. Jr.: L-structure in L-spaces. Trans. Amer. Math. Soc. **95**, 274—299, (1960).
412. — M-structure in Banach spaces. Proc. Cambridge Philos. Soc. **63**, 613—629 (1967).
413. Effros, E.: On a class of real Banach spaces (to appear).
414. Fakhoury, H.: Structures uniformes sur un cône bien coiffé. C. R. Acad. Sci. Paris **270**, 1365—1368 (1970).
415. — Structures uniformes sur une classe de cones convexes. C. R. Acad. Sci. Paris **271**, 266—270 (1970).
416. — Préduaux des L-espaces; propriétés des G-espaces et des C_σ-espaces. C. R. Acad. Sci. Paris **271**, 941—944 (1970).
417. — Espaces fortement réticulés de fonctions affines. Propriétés des simplexes dont l'ensemble de points extrémaux et $\mathcal{K}$-analytique. La Faculté des Sciences de Paris 1970 (mimeographed, pp. 68).
418. Gleit, A. S.: On the structure topology of simplex spaces. Pacific J. Math. **34**, 389—405 (1970).
419. — On the existence of simplex spaces. Israel J. Math. (to appear).
420. — A characterization of M-spaces in the class of separable simplex spaces (preprint).
421. Goullet de Rugy, A.: La topologie AA-faciale et son utilisation dans la théorie des cônes biréticulés. C. R. Acad. Sci. Paris **271**, 319—322 (1970).
422. — Précision sur la représentation des cônes bireticulés comme cônes de mesures. C. R. Acad. Sci. Paris **271**, 353—356 (1970).
423. Guichardet, A., Kastler, D.: Désintegration des etats quasi-invariants des C^*-algèbres (to appear).

424. Hirsberg, B.: A measure theoretic characterization of parallelband split faces and their connections with function spaces and algebras. Aarhus Univ. Various Publ. Series, no 16, 1970 (p. 32, mimeographed).
425. Hustad: A norm preserving complex Choquet theorem. Math. Scand (to appear).
426. Ionescu Tulcea, A-C.: Topics in the theory of lifting. Ergebnisse der Math., Vol. 48. Berlin-Heidelberg-New York: Springer 1969.
427. Kastler, D.: (cf. Guichardet, A., Kastler, D.).
428. Leger, C., Soury. P.: Le convexe topologique des probabilités sur un espace topologique. Inst. Henri Poincaré Paris 1970 (mimeographed, pp. 51).
429. Lindenstrauss, J.: Some aspects of the theory of Banach spaces. Advances in Math. **5** 159—180 (1970).
430. Lion, G.: Familles d'operateurs et frontièrs en théorie du potentiel. Ann. Inst. Fourier (Grenoble) **16**, 389—453 (1966).
431. McWilliams, R.D.: On the w^*-sequential closure of subspaces of Banach spaces. Portugal. Math. **22**, 209—214 (1963).
432. Rao, M.: Measurable selections of representing measures. Aarhus Univ., Preprint Series no. 24 (1969/70) (mimeographed, pp. 6.)
433. Rogalski, M.: Topologies faciales dans les convexes compacts; calcul fonctionel et décomposition spectrale dans le centre d'un espace $A(K)$. Seminaire de Choquet 9, no. 3, Paris 1969/70.
434. — Quelques problèmes concernant une charactérisation des simplexes. C.R. Acad. Sci. Paris **296**, 645—647 (1969).

Subject Index

45. Linnik: Ergodic Properties of Algebraic Fields. DM 44,—; US $ 12.10
46. Krull: Idealtheorie. DM 28,—; US $ 7.70
47. Nachbin: Topology on Spaces of Holomorphic Mappings. DM 18,—; US $ 5.00
48. A. Ionescu Tulcea/C. Ionescu Tulcea: Topics in the Theory of Lifting. DM 36,—; US $ 9.90
49. Hayes/Pauc: Derivation and Martingales. DM 48.—; US $ 13.20
50. Kahane: Séries de Fourier Absolument Convergentes. DM 44,—; US $ 12.10
51. Behnke/Thullen: Theorie der Funktionen mehrerer komplexer Veränderlichen. DM 48,—; US $ 13.20
52. Wilf: Finite Sections of Some Classical Inequalities. DM 28,—; US $ 7.70
53. Ramis: Sous-ensembles analytiques d'une variété banachique complexe. DM 36,—; US $ 9.90
54. Busemann: Recent Synthetic Differential Geometry. DM 32,—; US $ 8.80
55. Walter: Differential and Integral Inequalities. DM 74,—; US $ 20.40
56. Monna: Analyse non-archimédienne. DM 38,—; US $ 10.50
57. Alfsen: Compact Convex Sets and Boundary Integrals. DM 46,—; US $ 12.70
58. Greco/Salmon: Topics in $\mathfrak{m}$-Adic Topologies. DM 24,—; US $ 6.60
59. Medrano, de: Involutions on Manifolds. DM 36,—; US $ 9.90
60. Sakai: C^*-Algebras and W^*-Algebras. In preparation